T0332611

Second Edition

HIGH-
FREQUENCY
and
MICROWAVE
CIRCUIT DESIGN

Second Edition

HIGH-FREQUENCY and MICROWAVE CIRCUIT DESIGN

Charles Nelson

CRC Press
Taylor & Francis Group
Boca Raton London New York

CRC Press is an imprint of the
Taylor & Francis Group, an **informa** business

CRC Press
Taylor & Francis Group
6000 Broken Sound Parkway NW, Suite 300
Boca Raton, FL 33487-2742

© 2008 by Taylor & Francis Group, LLC
CRC Press is an imprint of Taylor & Francis Group, an Informa business

No claim to original U.S. Government works
Printed in the United States of America on acid-free paper
10 9 8 7 6 5 4 3 2 1

International Standard Book Number-13: 978-0-8493-7562-0 (Hardcover)

Library of Congress Cataloging-in-Publication Data

Nelson, Greg (Charles G.)
High-frequency and microwave circuit design / Charles Nelson.
p. cm.
Includes bibliographical references and index.
ISBN 978-0-8493-7562-0 (alk. paper)
1. Microwave circuits--Design and construction. 2. Electronic circuit design.
3. Modulation (Electronics) I. Title.

TK7876.N46 2008
621.381'32--dc22   2007020172

Visit the Taylor & Francis Web site at
http://www.taylorandfrancis.com

and the CRC Press Web site at
http://www.crcpress.com

Contents

Preface to Second Edition

The first edition of *High-Frequency and Microwave Circuit Design* was aimed at enabling graduating electrical engineering students to design stable amplifiers using potentially unstable transistors. The book was rather brief and assumed a better background in undergraduate topics than can still, at this date, often be depended upon. This author presented impedance matching using irises in waveguides and found that the length of WR 90 he brandished aloft was the first waveguide many had ever seen. Thus, it seemed advisable to add Chapter 2, "Waveguides," before bringing up irises. It was then reasonable to do impedance matching in Chapter 3.

The first edition covered stability circles in the first chapter. This part of the presentation made use of scattering coefficients, which are not commonly taught in any standard undergraduate texts on distributed circuits. We needed scattering coefficients even before they had been introduced. So, Chapter 4 introduces scattering coefficients and then examines the topic of stability. Chapters 5 and 6 were the original Chapters 3 and 4, on tuned circuits and oscillators and modulation circuits. Chapter 7 deals with the origin and control of noise powers. Chapter 8, on antennas, takes a somewhat different, qualitative approach than the previous edition; its exercises relate antenna performance to the requirements of digital communication systems. Throughout, the author has endeavored to offer practical homework problems.

Preface to the First Edition

"All right, Dr. Nelson," the potential reader may be thinking or saying, "what unfulfilled need does this book aim to fill?" Well, I say, every year we expect B.S. graduates in electronics engineering to have learned something that we formerly saved for graduate work. The material in this book is like that. The curriculum at this non-Ph.D.-granting institution requires a year of active circuit design from a book such as Microelectronic Circuits by Sedra and Smith. Thus, the student's exposure to high frequency effects in electronics will stop with the hybrid-pi high frequency equivalent circuit, leaving him or her to be most astonished and feeling helpless when they encounter a spec sheet which states scattering coefficients. Also, even though the curriculum might require an electromagnetics course treating wave behavior, the topics in electromagnetics books, even though they may treat standing waves and impedance matching, often do not mention that one's knowledge of the impedance to be matched may actually be stated in terms of scattering coefficients. Thus, a concise introduction to scattering coefficients that is understandable to undergraduates is a key aim of this book.

If there were one somewhat specialized audience at which this book is aimed, it would be those engineers who are or will be working to provide the world with the vast number and types of communication apparatus which will be needed at the beginning of the twenty-first century. These devices will have various digital functions and features, yet there will also need to be in them various aspects of analog design. The engineering graduate who thinks he or she will be able to confine herself, or himself, to either analog or digital design will quickly find that one has voluntarily limited one's usefulness to an employer. The author hopes to "demystify" some of the useful techniques that future communication engineers may need. He hopes that he can emphasize that many concepts have not changed and that the newer concepts and perspectives easily grew out of the traditional. One of his major aims has been not to startle the reader with many totally new or unfamiliar techniques.

Resonant circuits are a topic to which the student may have been introduced in both network analysis and active circuits courses. Yet, those presentations may both have been rather academic and totally lacking in relation to practical needs. The author hopes to motivate study of this topic by examples and exercises, which are applied to transmitters, receivers, and other applications requiring selectivity in the signals which are passed or rejected.

Chapter 1 begins at medium frequencies and quickly arrives at what are generally considered microwaves. It calculates the easily obtained distributed parameters in coaxial lines and determines their propagation constants and low and high frequency characteristic impedance. It moves on to the design of microstrip lines for specified characteristic impedance. Reflection coefficients and their effect on input impedances of lines are next derived, as well as the effects of attenuation on input impedance. Use of the Smith chart is reviewed. Finally, we arrive at scattering coefficients as two-port parameters for transistors. We are introduced to stability considerations and conditions for simultaneous conjugate impedance matching at input and output.

In Chapter 2, we begin the task of impedance matching by considering using L-sections of lumped impedance. We move on to quarter-wavelength and near-quarter-wavelength transmission lines for impedance matching; how and when we may use a matching section, which is not exactly a quarter-wavelength, is a straightforward procedure that is not commonly taught. Then we consider how one shunt or series reactance may match the transmission line if it is placed at the correct location. Finally, we see how the required reactance may be provided by an appropriate length of short or open-circuited transmission line, called a "stub." For completeness, we have also included irises for the matching of waveguides.

In Chapter 3, we begin with bandpass amplifiers using LRC circuit design and crystal filters. We apply these concepts to the needs of receivers and calculate performance limitations. We look at what are the requirements for a successful oscillator and illustrate it with perhaps conceptually the simplest oscillator, the phase shift oscillator. We move on to the increasing complexity of the Colpitts, the Hartley, and the crystal-controlled variety. Finally, we consider how one might design a voltage-controlled oscillator, as may be found in phase-locked loops.

Chapter 4 looks first at the most elementary ways of producing and detecting amplitude and frequency modulation. Synchronous detectors are the first demodulators considered, then later, the less expensive envelope detector. Finally, we look at the basics of the most common forms of digital modulation and demodulation.

Noise can be a very complex and theoretical topic. The approach here in Chapter 5 is simply to state the results using as given constants the arcane symbols of statistical mechanics. We define noise factor and noise figure in functional ways and look at the general way in which noise is added as we cascade amplifiers. Having established that overall noise performance depends mainly upon both the gain and the noise factor of the first stage, we learn how to draw amplifier noise and gain circles so as to have a systematic way to make such tradeoffs. We finish by seeing the high noise performance of FM and the markedly higher performance of systems of digital modulation.

Chapter 6 aims to lift, at least a little, the veil of mystery which has for many years hidden from young engineers some rather straightforward characteristics of antennas. The principles are introduced in such a way that a minimum amount of theory leads to a number of important performance characteristics. An aspect of this chapter as a capstone for the book is to use antenna gains, transmitter power, distance, carrier frequency, and bit rate in the evaluation of the optimum communication systems.

The computer revolution has also made available very powerful calculators which can do complex algebra, for reasonable prices. Appendix A gives a few helpful formulas which enable much of the impedance match problem to be soluble directly on the calculator.

Einstein was quoted that we should make complex things as "simple as possible, but no simpler." It is fondly hoped that this book can make the areas of high frequency design as clear as possible to the graduate electrical engineer.

About the Author

Charles G. Nelson, Ph.D., was born in Northport, Michigan. He received all his primary education in a tiny school district and graduated as co-valedictorian of his class. All of his academic degrees are in electrical engineering, including the bachelor of science from Michigan State University and the master of science and doctor of philosophy degrees from Stanford University. His doctoral research involved conversion efficiency of a klystron using varying profiles of magnetic field to hold the electron beam together. A summary was published in the Transactions on Electron Devices of IEEE. Other publications were in the Annual Convention on Engineering in Medicine and Biology and periodic and final reports to the California Department of Transportation (CALTRANS).

This is his first textbook to be published, although his students have for years used his photocopied notes as the only textbook for two or three subjects.

His military service was with the Research Lab of the Ordnance Missile Labs at Redstone Arsenal, Huntsville, AL, where his primary assignment was the study of a communication system using pulse position modulation with only discrete positions allowed, thus exhibiting some of the advantages and limitations of digitization. He had industrial experience with Zenith Radio Research Corporation of Menlo Park, CA, working on extending the operation frequency of the electron beam parametric amplifier, and summer research with NASA at Ames Research Labs, doing early studies of phase modulation of a light beam in lithium niobate. His research interests continue to be in fields and waves and communication systems, and in sound and noise pollution.

Dr. Nelson has taught electrical engineering at California State University, Sacramento since February 1965. He served as chair of the department of electrical and electronics engineering from 1965 to 1967 and from 1979 to 1986. He has been active in various areas of faculty governance for many years. He has been registered by the State of California as a Safety Engineer. He is a member of Tau Beta Pi, Pi Mu Epsilon, and Sigma Xi, and a life member of the IEEE.

Dr. Nelson has been married to Nina Volkert-Nelson since 1967. They have two adult children, a son educated as a computer scientist and a daughter who has her master's degree in cello performance. In his leisure, Dr. Nelson enjoys reading both fiction and nonfiction, from James Thurber to John LeCarre to Robert Woodward. He loves listening to a variety of instrumental and vocal music, is considered to be an opera buff, and he sings at the level of the better church choirs. He also enjoys the Pacific surf and wildlife and the tranquility of the shores of Lake Michigan.

1 From Lumped to Distributed Parameters

1.1 EXPECTED AND UNEXPECTED RESULTS FROM THE NETWORK ANALYZER

One very useful instrument for studying high-frequency behavior of circuits is called the *vector network analyzer*; an example of one is pictured in Figure 1.1. Measurements can be made of impedance connected to either port 1 or port 2, or of the amount of signal gain and phase shift from one port to the other, when an amplifying or attenuating two-port is connected between the ports. One first selects a frequency range of interest between 0.3 and 3000 Mhz, and then calibrates the instrument for that range and for the type of measurement planned. As an example of what is possible, the machine was calibrated from 0.3 to 100 Mhz for measurements of the impedance connected to port 1. After calibration, a type N-to-BNC coaxial adapter was connected to port 1, and a BNC-to-binding-post adapter was connected to the first adapter, after which a nominal 47 pF capacitor was connected to the binding posts without shortening the original capacitor leads. The reactances measured are shown in Table 1.1.

Now, these data are not too startling at the lower frequencies, although one observes a modest bit of jumping around. The next thing done was to disconnect the capacitor and make measurements of the impedances of just the adapters that were added after calibration. Again, the data did some jumping around, but we could say the adapters provided a fairly consistent 6.0 pF. Hence, since the capacitances of the adapters and the capacitor being measured are effectively in parallel, they add, and we can probably say with good accuracy that the "official" capacitor provides very close to 46 pF, which is certainly well within 10% of the nominal value. But now we ask, "What on earth is going on at high frequency?" and we must suspect that the long leads are having an effect. Indeed, the total negative reactance is decreasing at high frequency faster than it would for constant capacitance, until eventually the reactance goes positive, that is, inductive, so we must conclude that the long leads provide some inductance, the reactance of which cancels some of the capacitive reactance, giving a lower net reactance. Every little segment of wire contributes some inductance, so we say inductance is *distributed* uniformly along the leads. The "smart" but naive machine calculates a single equivalent circuit element, which is a capacitor higher than the nominal value, until one reaches the frequency where the inductive reactance in the leads takes over and makes the total impedance inductive.

1

FIGURE 1.1   Network analyzer with adapters and unknown capacitor connected.

TABLE 1.1
Reactances

Frequency (Mhz)	Reactance	Eq. Circuit Element
1.0	$-j3050$	52.2 pF
2.0	$-j1535$	51.8 pF
5.0	$-j606$	52.5 pF
10	$-j302$	52.7 pF
20	$-j144.6$	55.05 pF
50	$-j38.18$	83.3 pF
100	$+j21.6$	34.4 nH

Example 1.1

The author once heard and has for years been quoting a rule of thumb that one can expect component leads to inject a nanohenry of inductance for each millimeter of wire. Let's play with our data a bit and see how well this rule checks out. The reactances we obtained above were each the average of 16 individual measurements, so we will assume they were accurate and calculate the corresponding admittance (which is called *susceptance*). We will assume that the adapters add 6.0 pF and subtract that susceptance from the total. We convert the new susceptance back to impedance and subtract the reactance corresponding to 46.0 pF. We will attribute the difference to inductive reactance and calculate the lead inductance from that. We will do this at 20 MHz and leave the numbers at 50 and 100 MHz for the exercises.

Solution: $-1/j144.6 = j6.93 \times 10^{-3}$.

Now, 6.0 pF yields $Y = j7.54 \times 10^{-4}$ at 20 MHz. Subtracting gives $Y = j6.16 \times 10^{-3}$, or an impedance of $Z = -j162$. At 20 Mhz, 46.0 pF yields an impedance of $-j173$. Therefore, a subtraction gives the result that the inductance is adding $j11$ ohms, which at 20 Mhz yields an inductance of 87.5 nH. Since the lead lengths totaled about 55 mm, our result says to expect about 1.5 nH/mm of lead length.

Exercise 1.1
Repeat the calculations above for 50 and 100 Mhz. What inductance do you get?

Answers: You should get 87.9 and 86.5 nH, respectively.

The results above represent a departure from the type of circuit modeling the student of electrical engineering or electronics technology first sees, in which the circuit parameters are assumed, with good justification, to be "lumped" in a fairly obvious location. The capacitor in the example above appeared physically as a lump of dielectric material with leads sticking out of two ends. One is normally taught that these leads are basically short circuits, not contributing any resistance or other impedance. What we found out was that the leads contributed small amounts of inductance, but that at high frequencies the impedance of the inductance became higher than the impedance of the intended capacitance. Sometimes the frequency at which the reactances are equal is called the *frequency of self-resonance*.

1.2 INTERWIRE CAPACITANCE IN AN INDUCTOR

Once one begins thinking about accidental circuit parameters that may appear, one might wonder, "Suppose we set out with wire and a coil form to wind an inductor. Can we expect every bit of wire to contribute capacitance to every other bit of wire?" The answer is, "Yes, indeed." We will in fact provide a "recipe" for the RF (radio frequency) engineer to fabricate handmade inductors because this is one component that may commonly need to be connected to two terminals of an integrated circuit in which many functions have already been provided. The procedure is discussed in *Reference Data for Radio Engineers*, published in 1956 by the Howard Sams Co. We illustrate several aspects of such design in the following example.

Example 1.2   Winding a Coil for Specified Inductance
The author was hoping to wind a coil with inductance close to 0.5 μH. He took AWG #12 wire because he had some and wanted to keep resistance as low as possible, hence Q high. Also, the wire was stiff enough to avoid any accidental short circuits between adjacent turns of the wire. The temporary coil form was a ballpoint pen. Ten turns were wound. The resulting coil was 3 cm long with a diameter of 1 cm. The handbook mentioned in the paragraph above predicts for a solenoidal coil, an inductance given in microhenries by

$$L = n^2 dF,$$

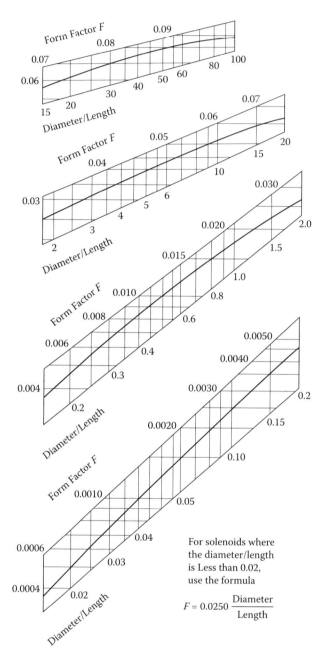

FIGURE 1.2   Form factor for single layer coil.

where n is the number of turns, d is the coil diameter in inches, and F is a *form factor*, depending upon the ratio of coil diameter to length, the form factor being given by the graph in Figure 1.2. The author read $F = 0.0075$ for the diameter to length ratio of 0.33, $d = 0.394$ inch, so for $n = 10$, the inductance should be 0.295 µH.

Network analyzer measurements at the lower frequencies were consistently 0.300 to 0.305 µH, so the procedure seems to work most satisfactorily.

Now, the next thing one might ask is, "What is the frequency of self-resonance and what capacitance does that represent?" From the network analyzer, self-resonance occurs at 119.3 Mhz. Capacitance to be accounted for is

$$C = \frac{10^6}{(2\pi \times 119.3 \times 10^6)^2 \times 0.3} \approx 6 \text{ pF}.$$

As can be seen from the previous example, the total capacitance can be accounted for by the capacitance of the adapters connected to the network analyzers. Thus, we have not really found the wire-to-wire capacitance. Some indication of what to expect may be obtained from Krauss, Bostian, and Raab. In their book, *Solid State Radio Engineering* (Wiley, New York, 1982), they give graphs of expected self-resonant frequencies for RF chokes (a somewhat generic name for inductors meant to have high impedance at high frequencies). For a 0.3 µH choke, they predict resonance between 200 and 450 Mhz. This would predict winding capacitance between 2.1 and 0.42 pF. The coil was not tightly wound, so the expected capacitance might be nearer the lower value, and it is not a great surprise that it apparently had no effect on the frequency of self-resonance.

Exercise 1.2
Predict the number of turns required for inductances of 0.5 µH and .1.0 µH, continuing with 1.0-cm diameter coils. However, estimate a new number of turns, assuming the coil length increases proportionately to the number of turns, and converges to a number. For 15 turns and a length of about 1.57 inches, the author finds $F = 0.0055$, $L = 0.501$ µH. Extending this to 1 µH, the calculation yields 22 turns and a length of 4.4 cm, 1.73 inch.

1.3 DISTRIBUTED PARAMETERS IN COAXIAL LINES

We have seen how distributed parameters can arise in rather simple circuits. Next we will see how they can become important in conductors, which at low frequencies are expected to be perfect conductors for connecting generators to loads without intervening impedance. We first compute the distributed parameters in coaxial lines; coaxial lines were chosen because the symmetry of their geometry permits accurate results without a lot of finagling and further simplifying assumptions. Later, we will make some generalizations relative to other geometries.

Consider the cross-sectional view of a coaxial line in Figure 1.3. To obtain capacitance, we need to relate the voltage

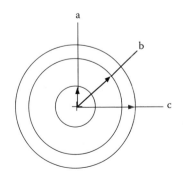

FIGURE 1.3 Geometry of coaxial line.

difference between conductors to the charge on the conductors. We assume the line to be effectively infinite in length, so that our calculation is true anywhere on the line. It is not currently feasible to talk about the total charge on one of the conductors, as it would also be infinite; instead, we assume that whatever charge is on a conductor is distributed uniformly along the length of the conductor with a density ρ_L Coulombs per meter. We apply Gauss's law over a cylinder having a as the radius of the inner conductor and b, the inside radius of the outer conductor. Since the Gaussian surface is of one meter length and has radius r, which is between a, and b, the radii of the inner and outer conductors, and is also coaxial with the conductors, the symmetry permits us to assume that the displacement D will be uniform over the Gaussian surface and will not have a non-zero z-component; hence, our closed surface integral has zero contribution from the ends of the cylinder, and, with a uniform D, the integral of D over the closed surface is simply the product of it and the area of the curved surface. We equate this to the charge enclosed by a meter-long cylinder and obtain

$$\oint D_r \bullet ds = D_r x 2\pi r(1) = \rho_L;$$

hence, $D_r = (\rho_L/2\pi r)$. To get the voltage between the conductors, we need to integrate

$$E_r = \frac{D_r}{\varepsilon},$$

so,

$$V = \int_a^b \frac{\rho_L dr}{2\pi\varepsilon r} = \frac{\rho_L}{2\pi\varepsilon} \ln\frac{b}{a}.$$

Since normal capacitance is the ratio of charge to voltage, here we find the ratio of *charge per meter* to voltage and obtain *farads per meter*. We will find that all of the charge will be distributed uniformly and will designate the capacitance per meter as

$$C_L = \frac{\rho_L}{V} = \frac{2\pi\varepsilon}{\ln\frac{b}{a}}.$$

The reader should remember that the quantity ε is the product of a *relative dielectric constant* ε_r for the dielectric between conductors times the so-called permittivity of vacuum ε_0. Since it is the only dimension in the capacitance expression that does not cancel out, clearly the units of ε_0 may be considered to be farads/meter.

Next, electromagnetics theory tells us that if there is a magnetic field intensity H resulting from currents in the conductors of a coaxial line, we can state a density of magnetic stored energy:

$$W_M = \frac{\mu}{2} H^2 \text{ Joules/cubic meter.}$$

Of course, the circuit element that stores magnetic energy is called an *inductor*, so next we look for the total energy stored between the conductors of a coaxial line. The simplest theory tells us that if there is a current in the center conductor, then an equal and opposite current flows in the outer conductor. We obtain H through the use of Ampere's law, which says that the line integral of H around a closed path gives the total current enclosed.

Since a path that is totally outside both conductors encloses no net current, we can infer that $H = 0$ outside both conductors; there is therefore a shielding of the "outside world" from the currents flowing in the conductors, as long as they are equal and opposite. We should emphasize that certain naive mistakes such as connecting an antenna of balanced geometry to the basically unbalanced geometry of a coaxial line without using a *balun* (a kind of transformer) can destroy the condition of equal and opposite currents and hence ruin the shielding effect.

Now, we can make clever use of Ampere's Law by applying it on a path that is coaxial with the inner and outer conductors, along which symmetry assures that the value of H is a constant, and the line integral is simply the product of the integrand and path length. So, when the path of integration has radius r, which is greater than a and less than b, we can say that

$$\oint H \bullet dl = 2\pi r H_\phi = I; \text{ therefore, } H_\phi = \frac{I}{2\pi r}$$

and the total energy stored between the conductors of a one meter length of coaxial line is the integral of energy density over the volume of the dielectric, as

$$W_m = \int_0^1 dz \int_0^{2\pi} d\phi \int_a^b \frac{\mu}{2} \left(\frac{I}{2\pi r} \right)^2 rdr = \pi\mu \left(\frac{I}{2\pi} \right)^2 \ln\frac{b}{a}.$$

We know of course that the energy stored in an inductor is

$$1/2 \ LI^2,$$

so equating this to the expression above, we get the inductance expression

$$L = \frac{\mu}{2\pi} \ln\frac{b}{a}$$

(in Henries/meter, as long as μ is in SI units [International System of Units]). This is sometimes called *external inductance* because it is related to energy stored in neither conductor, but between them.

There is of course non-zero magnetic intensity inside both conductors at low frequencies, but the so-called *skin effect* reduces such fields to insignificance at high frequency, and external inductance is the only significant inductance at high frequencies. We have mentioned before that there is a small penetration of currents into the conductor at high frequencies. The condition called skin effect is the phenomenon that, at high frequencies, current in a conductor crowds toward the surface of the conductor. If the conductor has two surfaces, as does the outer conductor of a coaxial cable, the current may be considered to flow in a thin layer on the inside of the outer conductor. The effective thickness of the shell of current is called the *skin depth*. The formula for skin depth is given as

$$\delta = \frac{1}{\sqrt{\pi f \mu \sigma}} \text{ meters,}$$

if all the electrical properties are given in SI units. Thus, we may calculate high-frequency resistance by assuming the current flows in layers one skin depth in thickness over the outside of the inner conductor and the outside of the outer conductor. Thus, high-frequency resistance of a coaxial line is given by

$$R_L = \frac{1}{2\pi\sigma\delta}\left(\frac{1}{a}+\frac{1}{b}\right) \text{ ohms/meter.}$$

We mentioned before the small penetration of currents into the conductor at high frequencies. Small currents lead to small amounts of magnetic energy, hence the resulting internal inductance is small. At high frequencies it may be calculated simply as $L_i = R/\omega$ and is normally very small compared to the external inductance derived above, but normal good practice might be to calculate it and justify neglecting it.

One more standard distributed circuit element is called *conductance* and results from the circumstance that standard dielectrics may have very small amounts of conductivity and other ways in which they may dissipate energy. Circuitwise, the best way to write this is as though the dielectric constant has a negative imaginary part in addition to its usually known real part. Thus, we may write permittivity with a real and imaginary part:

$$\varepsilon = \varepsilon_0 \varepsilon_r (1 - jd.f.)$$

Here the normal quoted permittivity of the dielectric is the quantities outside the parentheses and *d.f.* stands for *dissipation factor*, which may be quoted for several values of frequency in handbooks that give the properties of commonly used dielectrics. Therefore, the convenient way to compute a transmission line conductance is

$$G_L = \omega C_L x(d.f.).$$

Example 1.3

We have given a number of formulas without applying any of them, so now is a good time to put them to use. Let us calculate all the distributed circuit parameters at a frequency of 1 MHz for an RG 58U coaxial line. The "official" numbers are: inner conductor diameter $2a = 0.030$ inches; dielectric diameter (hence inner diameter of the outer conductor) is $2b = 0.105$ inches; the dielectric is stabilized polyethylene, for which the official dielectric constant is $\varepsilon_r = 2.26$; and the dissipation factor may be assumed to be 0.0002. Note that inches are not SI units, and may need to be converted to meters. Assume both conductors are copper, for which the conductivity is

$$\sigma = 5.8 \times 10^7 \, \text{S/m}.$$

Solution: Distributed capacitance is independent of frequency, and has the natural log of the *ratio* of dimensions, so we need not convert them to meters just yet:

$$C_L = \frac{2\pi \times 2.26 \times 8.854 \times 10^{-12}}{\ln(105/30)} = 100.4 \text{ pFd.}$$

Distributed external inductance is also independent of frequency, and is given by

$$L_e = \frac{\mu}{2\pi} \ln \frac{105}{30} = 250.6 \text{ nH/m.}$$

The skin depth in copper at 1 Mhz is

$$\delta = \frac{1}{\sqrt{\pi \times 10^6 \, x 4\pi \times 10^{-7} \times 5.8 \times 10^7}} = 6.608 \times 10^{-5} \text{ meters.}$$

In the distributed resistance formula, we have uncanceled dimensions, so we must be sure to convert them to SI by multiplying inches by 0.0254m/in. Also, one must suspiciously note that the dimensions were given as diameters, so a 2 must be inserted or divided out somewhere. Thus

$$R_L = \frac{1}{2\pi \times 6.608 \times 10^{-5} \times 5.8 \times 10^7 \times .0254} \left(\frac{2}{.030} + \frac{2}{.105} \right) = 0.140 \text{ ohms/meter.}$$

Now, it is elementary to compute internal inductance by dividing this result by radian frequency:

$$L_i = \frac{0.140}{2\pi \times 10^6} = 22.2 \text{ nH/m.}$$

As this is about 8.9% of the external inductance, the nature of the problem we are solving may govern whether we neglect it or not.

For a dielectric dissipation factor of 2×10^{-4}, the conductivity term becomes

$$G_L = 2\pi \times 10^6 \times 1.004 \times 10^{-10} \times 2 \times 10^{-4} = 1.26 \times 10^{-7} \text{ Siemens/meter.}$$

Exercise 1.3

Repeat the calculations in the example above at frequencies of 100 Mhz and 10 Ghz.

Answers: Capacitance and external inductance are unchanged. At 100 Mhz, $R_L =$ 1.4 ohms/meter; internal inductance becomes 2.2 nH/m, which, being less than 1% of the external inductance, is easy to neglect; and G_L becomes 1.26×10^{-5} S/m. At 10 Ghz, $R_L = 14.0$ ohms/m, $L_i = 0.22$ nH/m, and $G_L = 1.26 \times 10^{-3}$ S/m.

Exercise 1.4

Repeat the calculations of Example 1.3 and Exercise 1.3 for an RG 6U cable, which your friendly Radio Shack store is apt to sell you for connecting TV's or FM to an antenna or to the connector your savvy landlord provided in the living room of your apartment. The main differences between it and the cable in Exercise 1.3, which is used in a lot of electronics work, are dimensions and dielectrics. RG 6U has $2a = 0.040$ inches and $2b = 0.190$ inches. The dielectric of RG 6U is foamed polyethylene, used because it reduces attenuation of the cable to practical nonsignificance; its dielectric constant may be used as 1.49, and the author recommends the reader not worry about its dissipation factor or attempt computation of G_L.

Answers: $C_L = 53.2$ pF/m, $L_e = 311.6$ nH/m. $R_L = 0.0989$ ohms/m and $L_i = 15.75$ nH/m at 1 Mhz. $R_L = 0.989$ ohms/m and $L_e = 1.575$ nH/m at 100 Mhz. $R_L = 9.89$ ohms/m and $L_e = 0.16$ nH/M at 10 Ghz.

Exercise 1.5

The beleaguered reader may wonder if there is a practical application of these calculations. The answer is, yes, partly; we can see what could be a very bad idea. Suppose we had a video camera with an output impedance of 10 kΩ feeding a VCR with input impedance of 10 kΩ through 5 meters of RG 58U cable. Since video is what we call a *baseband signal*, meaning its low frequency is near zero, the equivalent circuit for this system simply has the two 10 kΩ in parallel with the cable capacitance, which will total around 500 pF. The reader should recognize a low-pass filter with half-power bandwidth given by

$$B = \frac{1}{2\pi RC}.$$

Find B.

Answer: 636 kHz. Putting the video through such a system may well strip off all the color information, which goes up to about 4 Mhz.

Exercise 1.6
The consequences of Exercise 1.5 illustrate the reason that emitter follower output is such a good idea for devices like video cameras. Suppose the only change you make is to get the camera output impedance down to 100Ω, which should not to be difficult at all. Find the new value of low-pass cutoff frequency *B*.

Answer: 31.9 Mhz, which ought to be high enough to pass video undiminished.

1.4 EQUIVALENT CIRCUIT FOR COAXIAL LINE

In the last section we found an expression for the capacitance for each meter of a coaxial line. The more line there is, the more capacitance there will be; since capacitors add in parallel, all of the capacitors try to be in parallel. Since conductance was calculated directly from capacitance, it is also a parallel element. Inductances add when they are put in series, as do resistors; hence, inductance and resistance are series elements of the equivalent circuit. The result is that the equivalent circuit for a meter or a really infinitesimal length of transmission line is as shown in Figure 1.4.

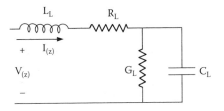

The first step in the analysis of this equivalent circuit is to write equations showing how voltage and current vary along the line. We will make the convenient assumption that we will always be expecting to excite the line sinusoidally, that is, the form of time expression for voltage would be

FIGURE 1.4 Equivalent circuit for an incremental length of transmission line.

$$v = V(z)e^{j\omega t},$$

where we expect the voltage to vary also as a function of z measured along the line, and where our objective is to determine this variation.

We can expect that the voltage will decrease as we go along the line, due to the voltage drop in L_L and R_L, so we can write

$$-\frac{\partial V}{\partial z} = R_L i + L_L \frac{\partial i}{\partial t} = R_L i + j\omega L_L i.$$

We can expect a similar drop in current as we go along the line, due to the current being shunted off through C_L and G_L. Hence, a second equation is

$$-\frac{\partial i}{\partial z} = G_L v + C_L \frac{\partial v}{\partial t} = G_L v + j\omega C_L v.$$

Heaviside called these the *telegrapher's equations*, not really recognizing that the man with a fast wrist sitting in a railroad station wearing a green eyeshade didn't have Heaviside's math skills, but we'll say having a name is better than not having one. For the moment, we will simply take a second z-derivative of the first equation and then substitute the second equation for $\partial i/\partial z$. We get

$$-\frac{d^2 v}{dz^2} = \left(R_L + j\omega L_L \right)\frac{di}{dz} = -\left(R_L + j\omega L_L \right)\left(G_L + j\omega C_L \right)v.$$

If we take the second derivative to the other side, we can say we have a second-order differential equation with zero as its first derivative, so it is of the form

$$\frac{d^2 v}{dz^2} - \gamma^2 v = 0.$$

The solutions to this equation are of the form $Ae^{-\gamma z} + Be^{\gamma z}$, where the quantity

$$\gamma = \sqrt{\left(R_L + j\omega L_L \right)\left(G_L + j\omega C_L \right)}.$$

Here two complex numbers are multiplied and the square root is taken. If the reader thinks he/she can avoid another complex number, the author has some swampland for sale. Typically, we define the parts $\gamma = \alpha + j\beta$. Since the real part α leads to a real exponential in v, we call α the *attenuation constant*. In a similar definition, β is called the *phase shift constant*. If we so chose, we could also define a *phase velocity*, U, for waves at this frequency as $U = \omega/\beta$; we would find that at frequencies where internal inductance can be neglected, waves travel at the velocity of light traveling in the dielectric used.

Example 1.4
Let us calculate the attenuation constant and phase shift constant in RG 58U cable at 1 MHz.

Solution: Using the numbers from Example 1.2, we first add the internal inductance result to the external inductance. We then have

$$\gamma = \sqrt{(0.140 + j2\pi \times 10^6 \times 2.72.6 \times 10^{-9})(1.26 \times 10^{-7} + 2\pi \times 10^6 \times 100.4 \times 10^{-12})}.$$

We show a few intermediate results in this calculation to put to rest a few possible reader impressions of magic:

$$\gamma = \sqrt{(0.140 + j1.713)(1.26 \times 10^{-7} + 6.308 \times 10^{-4})}$$

$$= \sqrt{1.719\angle 85.33° \times 6.308\angle 89.99°}$$

$$= \sqrt{1.084 \times 10^{-3}\angle 175.32°} = 3.29 \times 10^{-2}\angle 87.66° = 0.001345 + j0.03289.$$

The units here are nepers/meter for α and radians/meter for β. The consequences are that a wave must travel 743 meters before its amplitude drops to $1/e$ times the value at which it started. The phase velocity here is

$$U = 2\pi \times 10^6 / 0.03289 = 1.91 \times 10^8 \text{ meters/second.}$$

Thus we see that internal inductance has slowed the wave velocity about 4.5% below the velocity of plane waves in the dielectric, which is

$$U = \frac{3 \times 10^8}{\sqrt{\varepsilon_r}} = \frac{3 \times 10^8}{\sqrt{2.26}} = 1.996 \times 10^8 \text{ m/s.}$$

Exercise 1.7
For RG 58U, find the attenuation constant and phase velocity at 100 MHz and 10 GHz.

Answers: At 100 MHz, $\alpha = 0.01426$ nepers/m and $U = 1.985 \times 10^8$ m/s; at 10 GHz, $\alpha = 0.172$ nepers/m and $U = 1.993 \times 10^8$ m/s.

Exercise 1.8
For an RG 6U cable, find the attenuation constant and phase velocity at 1 MHz, 100 MHz, and 10 GHz.

Answers: At 1 MHz, 6.30×10^{-4} neper/m and $U = 2.395 \times 10^8$ m/s; at 100 MHz, 6.44×10^{-3} nepers/m and $U = 2.45 \times 10^8$ m/s; at 10 MHz, 6.46×10^{-2} nepers/m and $U = 2.455 \times 10^8$ m/s.

1.5 CHARACTERISTIC IMPEDANCE ON COAXIAL LINES

The ratio between voltage and current for the waves on a transmission line is defined as the *characteristic impedance*. First, we might say that we recognize the individual waves from their mathematical representation. We have been saying that the time representation is given by the imaginary exponential $e^{j\omega t}$. Now, we have two possible solutions to our wave equation; because we took a square root, they are the positive and negative of the same quantities. When we combine these parts with the time variation, the two waves may be written mathematically as

$$v = A e^{j\omega t} e^{-\alpha z} e^{-j\beta z} + B e^{j\omega t} e^{\alpha z} e^{j\beta z}.$$

The imaginary exponents with the coefficient A are $\omega t - \beta z$. Suppose we pick the constant equal to 0; we would say that at time $t = 0$, that phase of the wave is located at $z = 0$. As time goes on, t becomes increasingly positive, so to follow the phase equal to 0 point on the wave, we must let z increase. Hence, this is a wave traveling in the +z-direction. Similar reasoning shows that the wave with the coefficient B is traveling in the −z-direction. If we had substituted differently, we would have found that current on the coaxial line satisfied the same equation as was satisfied by voltage; hence, the solutions for current would be the same as those for voltage, except that the exponential functions would be multiplied by different constants. Let us take the positive-going voltage and insert it in the first telegrapher's equation. We have

$$-\frac{\partial v}{\partial z} = -e^{j\omega t}(-\alpha - j\beta) = \gamma v_{+wave} = (R_L + j\omega L_L)i_{+wave}.$$

Now, defining characteristic impedance as the ratio of voltage to current for the + wave gives us

$$Z_0 = \frac{v_{+wave}}{i_{+wave}} = \frac{R_L + j\omega L_L}{\gamma} = \frac{R_L + j\omega L_L}{\sqrt{(R_L + j\omega L_L)(G_L + j\omega C_L)}} = \sqrt{\frac{R_L + j\omega L_L}{(G_L + j\omega C_L)}}.$$

If we perform these same manipulations for the negative-going wave, we get

$$\frac{v_{-wave}}{i_{-wave}} = -Z_0.$$

We can say in general that characteristic impedance is a function of frequency. However, at high frequencies, we can say that $\omega L \gg R_L$, and for good dielectrics, $\omega C_L \gg G_L$. Hence, we may write

$$Z_0 \cong \sqrt{\frac{j\omega L_L}{j\omega C_L}} = \sqrt{\frac{L_L}{C_L}}.$$

Indeed, it is a normal approximation that the final expression is stated as the characteristic impedance of a cable. And, reiterating, for the wave going in the $-z$-direction, the ratio of v_{-wave} to i_{-wave} is $-Z_0$. The significance of the negative sign lies in the fact that we define positive current as flowing into the positive voltage conductor; the negative characteristic impedance implies only that the current flows *out* of the positive terminal.

Example 1.5
Find the exact characteristic impedance of RG 58U cable at 1 MHz and compare it to the usual approximation.

Solution:

$$Z_0 = \sqrt{\frac{0.140 + j2\pi \times 10^6 \times 272.8 \times 10^{-9}}{1.26 \times 10^{-7} + 2\pi \times 10^6 \times 100.4 \times 10^{-12}}}$$

$$= \sqrt{\frac{1.720\angle 85.33°}{0.0006308\angle 89.99}} = 52.2\angle -2.33°.$$

This is already very close to the approximate value

$$Z_0 = \sqrt{\frac{250.6 \times 10^{-9}}{100.4 \times 10^{-12}}} = 49.96.$$

In fact, the microwave engineer will almost invariably refer to RG 58U as an example of a 50-ohm cable.

Exercise 1.9
Find the exact value of the characteristic impedance of RG 58U cable at frequencies of 100 MHz and 10 GHz.

Answers: $50.18\angle - 0.025°, 49.96\angle - 0.20°$.

Exercise 1.10
Find the exact value of the characteristic impedance of RG 6U cable at frequencies of 1 MHz, 100 MHz, and 10 GHz.

Answers: $78.49\angle - 1.38°, 76.72\angle - 0.14°, 76.56\angle - 0.014°$.

This cable thus has a nominal characteristic impedance of 75 ohms, which is a good match for the folded monopole antenna, or to the 300 ohms of the folded dipole antenna when one also uses the impedance-matching transformer called a *balun*.

1.6 TERMINAL CONDITIONS—REFLECTION COEFFICIENT

On sinusoidally excited lines, the most interesting phenomena occur at or near the load end. Consequently, it is convenient to define waves in terms of their values at the load terminals. Figure 1.5 illustrates where the variables are valid, at $Ae^{-\gamma L} = V^+$ and $Be^{\gamma L} = V^-$, where we may say V^+ is the value *at the load* of the voltage wave traveling *toward* the load, and V^- is the value at the load of he wave traveling *away from* the load. Similarly, the values of current waves at the load are I^+ and I^-. But since voltage and current in each wave are related through characteristic impedance, we may write

$$I^+ = \frac{V^+}{Z_0} \text{ and } I^- = -\frac{V^-}{Z_0}.$$

Now, the load has a voltage–current relationship given by

$$Z_t = \frac{V_t}{I} = \frac{V^+ + V^-}{I^+ + I^-} = \frac{V^+ + V^-}{\dfrac{V^+}{Z_0} - \dfrac{V^-}{Z_0}}.$$

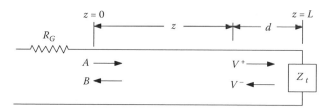

FIGURE 1.5   Direction of waves near the load.

If we now define a reflection coefficient as

$$\Gamma_T = \frac{V^-}{V^+},$$

we can obtain its relation to load impedance. We have

$$Z_T = \frac{V^+ + \Gamma_T V^+}{\dfrac{V^+}{Z_0} - \dfrac{\Gamma_T V^+}{Z_0}} = Z_0 \frac{1 + \Gamma_T}{1 - \Gamma_T}.$$

Solving this for Γ_T, we get

$$\Gamma_T = \frac{\dfrac{Z_T}{Z_0} - 1}{\dfrac{Z_T}{Z_0} + 1}.$$

When we divide an impedance by Z_0, we call the result *normalized impedance*, which is often denoted by a lowercase letter; thus

$$z_t = \frac{Z_T}{Z_0}.$$

This will be found to be an especially convenient concept later when we work with the Smith chart.

Exercise 1.11
Find Γ_T, for $Z_T/Z_0 = 2,, 0.5, \infty, 0, j, -j$.

Answers: $1/3, -1/3, 1, -1, 1\angle 90°, 1\angle{-}90°$.

1.6.1   GENERALIZED REFLECTION COEFFICIENT

At this point we define a *generalized reflection coefficient*, which is the ratio of the two voltage waves at any point on the line. It is particularly useful to handle loss on a transmission line by computing its effect on this generalized reflection coefficient. It is also especially convenient because modern vector network analyzers commonly read out this variable. Let us rewrite the voltage on the line in terms of the values of the waves at the load end. Solving for A and B in terms of the length of the line, which we could, at least momentarily, for the purpose of the derivation call L, we would get

$$A = V^+ e^{\gamma L} \text{ and } B = V^- e^{-\gamma L}, \text{ so}$$

$$V = V^+ e^{\gamma L} e^{-\gamma z} + V^- e^{-\gamma L} e^{\gamma z} = V^+ e^{\gamma(L+z)} + V^- e^{-\gamma(L-z)}.$$

The quantity $L - z$ measures how far a given point on the transmission line is from the load terminals, and is a useful quantity with which to work in many sinusoidal steady-state problems. We will call it d and define the generalized reflection coefficient as

$$\Gamma(d) = \frac{V^- e^{-\gamma d}}{V^+ e^{\gamma d}} = \Gamma_t e^{-2\gamma d} = \Gamma_t e^{-2\alpha d} e^{-j2\beta d}.$$

Since Γ_T and γ are both in general complex, we can see that the effect of moving away from the load is to retard the phase and shrink the amplitude of Γ. The latter effect can be quite salutary in producing a better impedance match at the generator when the load impedance alone would produce a severe mismatch. Once the generalized reflection coefficient has been defined, it is a big help for future calculations to express the general solutions for voltage and current on the line in terms of $\Gamma(d)$. Factoring $V^+ e^{\gamma d}$ out of the voltage expression, we get

$$V(d) = V^+ e^{\gamma d}\left[1 + \frac{V^- e^{-\gamma d}}{V^+ e^{\gamma d}}\right] = V^+ e^{\gamma d}(1 + \Gamma_T e^{-2\gamma d}) = V^+ e^{\gamma d}(1 + \Gamma(d)).$$

A similar manipulation can be performed upon the general solution for current on the line, obtaining

$$I(d) = \frac{V^+ e^{\gamma d}(1 - \Gamma(d))}{Z_0}.$$

These expressions for voltage and current are especially useful if one wishes to know what the input impedance of a line would be if it were cut at some distance d from the load. One simply divides the voltage expression by the current expression, obtaining

$$Z(d) = \frac{V^+ e^{\gamma d}(1 + \Gamma(d))}{\dfrac{V^+ e^{\gamma d}(1 - \Gamma(d))}{Z_0}} = Z_0 \frac{1 + \Gamma(d)}{1 - \Gamma(d)}.$$

Example 1.5

Given $\Gamma_T = 0.3 \angle 45°$, what will be the reflection coefficient at the input end of a transmission line 0.3 λ long with a one-way attenuation of 2 dB? If the characteristic impedance of the line is 50 ohms, what termination impedance caused the reflection coefficient and what is the input impedance of the line?

Solution: We need to evaluate $\Gamma_T = 0.3 \angle 45°$. Since $e^{-\alpha L}$ tells how much a voltage wave is attenuated in traveling one way on the line, we must say

$$-2 \text{ dB} = 20 \log_{10}(e^{-\alpha L}) = 10 \log_{10}(e^{-2\alpha L}); \ e^{-2\alpha L} = 10^{-0.2} = 0.631.$$

A more convenient way of expressing β is 2π/λ. Then

$$2\beta L = 2\left(\frac{2\pi}{\lambda}\right)0.3\lambda = 1.2\pi = 216°$$

$$\Gamma(L = 0.3\lambda) = 0.3 \times 0.631 \angle(45° - 216°) = 0.189 \angle -171°.$$

The expression for $Z(d)$ may be evaluated for any value of d, including zero. Hence,

$$Z_t = Z_0 \frac{1 + \Gamma_T}{1 - \Gamma_T} = 50 \frac{1 + 0.3 \angle 45°}{1 - 0.3 \angle 45°} = 50 \frac{1 + 0.2121 + j0.2121}{1 - 0.2121 - j0.2121}$$

$$= 50 \frac{1.2306 \angle 0.93°}{0.8159 \angle -15.069°} = 75.41 \angle 25.00° = (67.1 + j31.3)\Omega.$$

Similarly,

$$Z_T = Z_{0-} \frac{1 + \Gamma(d)}{1 - \Gamma(d)} = 50 \frac{1 + 0.189 \angle -171°}{1 - 0.189 \angle -171°} = 50 \frac{1 - 0.1870 - j0.0296}{1 + 0.01870 + j0.0296}$$

$$= 34.26 \angle -3.51° = 34.20 - j2.10.$$

Exercise 1.12

Given $\Gamma_T = 0.80 \angle 30°$, what is $\Gamma(L)$ at 4.25 λ away from the load on a line on which the one-way attenuation is 3 dB? If the characteristic impedance of the line is 75 ohms, what termination impedance caused the reflection coefficient and what is the input impedance of the line?

Answers: $0.4 \angle 210°, Z_T = (106.15 + j235.9)\Omega,, Z(d) = (34.00 - j16.19)\Omega.$

Exercise 1.13

The effect of attenuation on the generalized reflection coefficient suggests an effective method for determining the attenuation on a line. In principle, one connects a known reflection coefficient to the load terminals and measures the generalized reflection coefficient at the other end. In practice, the most convenient reflection coefficient to provide is −1, corresponding to a short circuit. Suppose a 50-ohm line is shorted and the input impedance is found to be 350 ohms. Find the one-way attenuation of the line as a decimal fraction and in decibels.

Answer: $e^{-\alpha L} = 0.866 \rightarrow -1.25$ dB.

Exercise 1.14

Repeat Exercise 1.13 for a shorted 50-ohm line with an input impedance of 3 ohms.

Answer: $e^{-2\alpha L} = 0.887 \rightarrow -0.52$ dB.

Exercise 1.15

Suppose input impedance is 5 ohms when a certain 75-ohm line is shorted. What is its attenuation factor?

Answer: $e^{-\alpha L} = 0.935 \rightarrow -0.58$ dB.

1.7 MR. SMITH'S INVALUABLE CHART

In many problems, as in Exercise 1.15 above, we see that neglecting attenuation totally would not greatly affect the calculated results. Even before World War II, a Bell Labs scientist named P. H. Smith derived a graphical aid to calculation that greatly simplified the calculation of input impedance, especially for lossless lines. On lossless lines, the generalized reflection coefficient keeps a constant amplitude, varying only in phase as one travels along the line; hence, the locus of the generalized reflection coefficient is a circle of radius determined by the amplitude of the coefficient. Smith might have said, "Consider the generalized reflection coefficient to be a complex variable with magnitude less than or equal to unity. We will write the variable as $u + jv$." Also, let us divide the input impedance by the characteristic impedance Z_0. The result is called *normalized impedance*; from here forward, we will use lowercase letters for normalized impedance or admittance. Hence, $z = r + jx$. Thus, we write normalized impedance as

$$\frac{Z(d)}{Z_0} = r + jx = \frac{1+u+jv}{1-(u+jv)}.$$

We need to separate the right-hand side into real and imaginary parts, which we will equate to r and x. This is done by multiplying by the complex conjugate of the denominator. Thus

$$r + jx = \frac{1+u+jv}{1-u-jv} \times \frac{1-u+jv}{1-u+jv} = \frac{1-u^2-v^2+j2v}{(1-u)^2+v^2}.$$

Hence,

$$r = \frac{1-u^2-v^2}{u^2-2u+1+v^2}.$$

Multiplying through to clear the denominator, we get

$$u^2 r - 2ur + r + v^2 r = 1 - u^2 - v^2.$$

The next operation we need to do is called *completing the square* in u and v. First we move several terms, getting $u^2(r + 1) - 2ur + v^2(r + 1) = 1 - r$. Dividing through by $r + 1$, we get

$$u^2 - \frac{2ur}{r+1} + v^2 = \frac{1-r}{1+r}.$$

Here the square is already completed in v, but the second term has u times "something," and we must simply add the square of one-half the "something," which is $r/(r + 1)$, to both sides of the equation. So,

$$u^2 - \frac{2ur}{r+1} + \left(\frac{r}{r+1}\right)^2 + v^2 = \left(u - \frac{r}{r+1}\right)^2 + v^2 = \frac{1-r}{1+r} + \left(\frac{r}{r+1}\right)^2$$

$$= \frac{1-r^2+r^2}{(1+r)^2} = \left(\frac{1}{r+1}\right)^2.$$

Now, our long-forgotten knowledge of analytic geometry will tell us that we have a family of circles with centers at $u = r/(r + 1)$ and $v = 0$, of radius $1/(r + 1)$. Let us just do a couple of these:

1. For $r = 0$, the circle is centered at $u = 0$, $v = 0$, that is, the origin, and of radius 1, which is a circle of unit radius centered at the origin.
2. For $r = 1$, the center is at 1/2, 0, and of radius 1/2. For $r = 3$, the center is at 3/4, 0, and of radius 1/4.

It can be seen that all r-circles go through the point (1,0). The equation for x almost has its square complete as we start

$$x = \frac{2v}{(1-u)^2+v^2}.$$

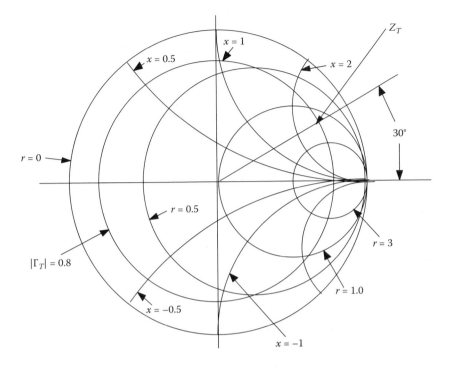

FIGURE 1.6   Smith chart constructed from discussion in text.

Hence

$$(u-1)^2 + v^2 = \frac{2v}{x}.$$

The square is completed in u, and we complete it in v by bringing $2v/x$ onto the left-hand side and adding the $(1/x)^2$ to both sides, obtaining

$$(u-1)^2 + \left(v - \frac{1}{x}\right)^2 = \left(\frac{1}{x}\right)^2.$$

Thus, circles of constant x are centered at $u = 1$, $v = 1/x$ and have radius $1/x$. In our sketch, Figure 1.6, we have shown circles for $x = 0$, $x = \pm1/2$ and ±1. Copies of the chart with many circles of both r and x are available at a nominal price in the bookstores of most colleges offering engineering or electronics technology; they may also, at this writing, be accessed on the Internet.

Example 1.6
Once again, if Candide were running the world, Smith charts would be perfectly round and precise in all ways. The author's experience is that some unnamed villains are subjecting good, accurate Smith charts to photocopy machines, which have

a slight tendency to stretch the image a bit vertically, so that what was once perfectly round is now slightly elliptical. We will act like Candide, and assume that the $\Gamma = 1$ ($r = 0$) circle on the Smith chart is precisely 91.2 mm in radius, but will give answers not from graphical calculations but from numerical ones, as are demonstrated in Appendix A. Suppose we wish to know what normalized impedance corresponds to a reflection coefficient of $1/3\angle 0°$?

Solution: The magnitude of this reflection coefficient will be given by one-third of 91.2 mm, or 30.4 mm. From the center of the chart, we measure 30.4 mm horizontally right, and if we have a perfectly round chart, we should find $z = 2.0 + j0$.

Exercise 1.16
Using the Smith chart, find the normalized impedances corresponding to 0.333 at 90, 180, and 270 degrees.

Answers: $0.8 + j0.6$, 0.5, $0.8 - j0.6$.

Exercise 1.17
Find the normalized impedances corresponding to a reflection coefficient magnitude of 0.5 at angles of 0, 90, 180, and 270 degrees.

Answers: 3.0, $0.6 + j0.8$, 0.33, $0.6 - j0.8$.

Exercise 1.18
Find the normalized impedances corresponding to a reflection coefficient $0.2\angle 30°$, $0.3\angle{-}40°$, $0.4\angle{-}110°$, and $0.5\angle{-}270°$.

Answers: $1.38 + j0.29$, $1.44 - j0.61$, $0.59 - j0.52$, $0.6 + j0.8$.

Exercise 1.19
Find the reflection coefficients if the following impedances are connected to a 50-ohm line: $10 + j20$, $75 - j25$, $100 - j200$, $-j50$, $j100$.

Answers: $0.71\angle 135°$, $0.28\angle{-}33.7°$, $0.82\angle{-}22.8°$, $1\angle{-}90°$, $1\angle 53.1°$.

1.8   CHARACTERISTIC IMPEDANCE OF MICROSTRIP LINE

We did a distributed circuit analysis of a coaxial line because, (1) historically, it has been very important to RF engineers, and (2) its simple geometry leads to fairly simple mathematics. Another kind of line has become fairly common because it connects rather conveniently to integrated circuit board electronics. It is called a *microstrip*, and its geometry is shown in Figure 1.7.

The analysis of a microstrip can be simple, but only when its dimensions permit neglecting fringing of electric fields outside the region between the plates, that is, when the spacing h is small compared to the width W of the signal conductor. Unfortunately, such conditions would limit one to rather low characteristic impedances,

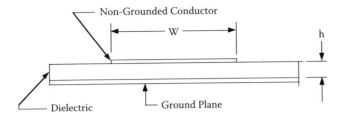

FIGURE 1.7   Geometry of a microstrip circuit.

and today's microwave engineer requires the flexibility of a variety of characteristic impedances. Our procedure will be to perform the approximate analysis that is accurate when width is large compared to spacing; for narrow conductor widths, we will simply use the fairly messy formula that a more precise analysis yields. When the approximation $W/h \gg 1$ is valid and fringing of electric fields can be neglected, we can recall elementary physics and simply write the capacitance per meter of distance into or out of the page as

$$C_L = \frac{\varepsilon_r \varepsilon_0 W}{h} \quad \text{Farads/meter.}$$

Next, we recall the results of coaxial cable analysis, in which it was found that at high frequencies, inductance is mainly the external inductance, which is very reminiscent of the capacitance formula, with ε replaced by μ, and reciprocals of all dimensions; thus

$$L_L = L_e = \frac{\mu_r \mu_0 h}{W} \quad \text{Henries/meter,}$$

where μ_r is the relative permeability of the dielectric and is close to unity for practical dielectrics.

Thus, when $W/h \gg 1$, high-frequency characteristic impedance is well approximated as

$$Z_0 \approx \sqrt{\frac{L_e}{C_L}} = \sqrt{\frac{\mu_0 h}{W} \times \frac{h}{\varepsilon_r \varepsilon_0 W}} = \frac{1}{\varepsilon_r} \sqrt{\frac{\mu_0}{\varepsilon_0}} \frac{h}{W}.$$

Suppose we evaluate this impedance for $\varepsilon_r = 2.25$, which is close to the value for one grade of Duroid, a product of Rogers Corporation of Chandler, AZ, and for $W = 10\ h$. We note that the quantity

$$\sqrt{\frac{\mu_0}{\varepsilon_0}}$$

is known in wave theory as the *intrinsic impedance of vacuum*, and is approximated by 120π ohms. Thus, the characteristic impedance is

$$Z_0 \approx 8\ \pi, \text{ about 25 ohms.}$$

The exact solution for the width of the conductor versus spacing for a given design value of characteristic impedance is complicated by its being a *two-dielectric problem*. The definitive word comes from Harold Wheeler in an Institute of Electrical and Electronics Engineers (IEEE) article published in 1977,* the year in which he turned 74 years of age. This writer judged the more useful result to be W/h for a specified characteristic impedance. If it is assumed that the thickness of the conductor is also small compared with the spacing, the formula becomes

$$\frac{W}{h} = 8\frac{\sqrt{\left[\exp\dfrac{Z_0}{42.4}\sqrt{\varepsilon_r+1}-1\right]\dfrac{7+4/\varepsilon_r}{11}+\dfrac{1+1/\varepsilon_r}{.81}}}{\left[\exp\dfrac{Z_0}{42.4}\sqrt{\varepsilon_r+1}-1\right]}.$$

Another consequence of having two different dielectrics is that waves travel on the line at a speed that is a complicated function of the dielectric constant and of the ratio W/h. The speed of waves is given by $c/\sqrt{\varepsilon_{eff}}$, where c is the speed of plane waves in vacuum and ε_{eff} is the effective value of the dielectric constant given by

$$\varepsilon_{eff} = \frac{\varepsilon_r+1}{2} + \frac{\varepsilon_r-1}{2}\left(1+\frac{12h}{W}\right)^{-1/2}.$$

Example 1.7

Let us suppose that we have microstrip material in which the actual dielectric constant is $\varepsilon_r = 2.25$ and the thickness h of the dielectric is 1.0 mm. For a characteristic impedance of 50 ohms, we wish to calculate the width W of the nongrounded conductor and the wave at a frequency of 1000 Mhz (1.0 Ghz). (Later we will study impedance-matching techniques that involve using lengths of transmission line that must be expressed in terms of wavelength.)

Solution:

$$\frac{W}{h} = 8\frac{\sqrt{\left[\exp\dfrac{50}{42.4}\sqrt{2.25+1}-1\right]\dfrac{7+4/2.25}{11}+\dfrac{1+1/2.2}{.81}}}{\left[\exp\dfrac{50}{42.4}\sqrt{2.25+1}-1\right]}.$$

If the reader wishes to check intermediate results, the author's result for the denominator (which also appears as a multiplier in the numerator) is 7.380, and the overall result is 3.00, or $W = 3$ mm.

---

* Wheeler, H.A. "Transmission Line of a Strip on a Dielectric Sheet on a Plane," *IEEE Transactions on Microwave Theory and Techniques* 25 (1977): 631–647.

Then,

$$\varepsilon_{eff} = \frac{2.25+1}{2} + \frac{2.25-1}{2}(1+12/3.0)^{-1/2} = 1.905.$$

Thus, the wavelength, which would be 30 cm in vacuum and 20 cm in a one-dielectric system completely insulated by its dielectric, would here be

$$\lambda = \frac{30}{\sqrt{1.90}} \text{ cm} = 21.74 \text{ cm}.$$

Exercise 1.20

For dielectrics of dielectric constants 3.27, 4.5, 6.15, and 9.8, find the value of W/h for a 50-ohm characteristic impedance, the effective dielectric constant, and the wavelength at 1 Ghz.

Answers: The answers are shown in Table 1.2, as a lesson to computer-phobics like the author. Even for us, such elementary ideas as spreadsheets can give superior results, in far less time, than the most powerful calculators, unless perhaps the latter are also programmed.

TABLE 1.2
Answers for Exercise 1.20

ε_r	Denominator in Formula	W/h	ε_{eff}	λ
3.27	10.436	2.352	2.594	18.625
4.5	14.888	1.876	3.393	16.286
6.15	22.411	1.473	4.426	14.259
9.8	47.202	0.976	6.607	11.671

2 Waveguides

2.1 INTRODUCTION

The microwave part of the electromagnetic spectrum received its major impetus for development before and during World War II; both sides feverishly pursued powerful radars that could detect the distance and direction of incoming bombers. Ground-based radars were also used in the fire-control systems of antiaircraft batteries. As soon as they could be shrunk in physical size, systems were also mounted aboard fighter planes, which could independently transmit pulsed signals and receive echoes from incoming bombers; they could thus search out and shoot down bombers in daylight or moonless night. Those wishing to read a fascinating history of the development of radar before, during, and after the war are referred to a book by Robert Buderi, *The Invention That Changed the World* (Simon and Schuster, 1996). As is true of much of history, the book was written by the winners. Nevertheless winners and losers have both benefited from radar, which tells them how to direct their airliners to avoid unfriendly weather. The whole world has also benefited from spin-offs such as the invention of the transistor, which permitted incredible miniaturization of electronic devices, making possible such other inventions as cell phones, space communication devices used on several trips to the moon and far beyond, and the little car that runs around on the Martian surface scooping up sand for examination.

In radar, a high-powered pulse of power (perhaps thousands of watts) is transmitted in an extremely narrow beam from a highly directive antenna. An extremely small fraction of the power is reflected from the target. To deliver large amounts of power from a transmitter tube to the antenna requires an extremely high-power handling transmission line. The coaxial cables used as laboratory transmission lines would be subject to arcing and short circuits if subjected to peak powers of kilowatts. Very early in the game, scientists and engineers realized that the only feasible transmission line was a waveguide. Physically, a waveguide is a pipe having a rectangular, cylindrical, or elliptical cross-section. It was quickly established that rectangular was the most practical shape.

2.2 FIELDS IN RECTANGULAR WAVEGUIDES

Figure 2.1 shows the geometry of a typical straight section of a rectangular waveguide. The convention that has arisen is to define the direction of wave transmission as the z-direction of a standard rectangular coordinate system, with the origin at one interior corner of the pipe. The longer of the transverse directions is called the x-direction and the shorter is the y-direction. *Interior* dimensions of the pipe are defined as a in the x-direction and b in the y-direction.

As emphasized in Chapter 1, signals can be sent along a two-conductor transmission line at any frequency from zero upward. Waveguide is a *one-conductor medium*

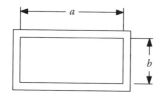

FIGURE 2.1 Rectangular waveguide showing inside dimensions.

in which waves travel along the inside of a pipe. The consequence is that transmission is a high-pass phenomenon. Next we will illustrate the mathematics demonstrating the conditions under which waveguides can transmit signals. We will seek a wave propagation equation for the dielectric region enclosed. We may expect the dielectric to be air at standard temperature and pressure, although this writer has experience using air at two or three atmospheres of pressure and with modest evacuation. (The infamous Mr. Murphy seems to have decreed that the *poorest* power handling in waveguide occurs right around one atmosphere, with improved performance at either higher or lower pressures.)

Our procedure is to start with Maxwell's Equations in differential form, with terms appropriate for a good dielectric. In general, Ampere's Law must admit the possibility that the medium must have finite conductivity, and that an electrical charge may be present. We can usually safely neglect such complications, so our form of Ampere's Law contains only a displacement current, and for a single sinusoidal signal, it thus becomes

$$\nabla \times \vec{H} = \varepsilon \frac{\partial \vec{E}}{\partial t} = j\omega\varepsilon\vec{E}.$$

Faraday's Law in this medium, again for a sinusoidal time variation, is

$$\nabla \times \vec{E} = -\mu \frac{\partial \vec{H}}{\partial t} = -j\omega\mu\vec{H}.$$

Here we can say we have two equations with the same two unknowns; a wave equation can be obtained by substituting from one equation into the other. Suppose we take a second curl in Faraday's Law. We get

$$\nabla \times \nabla \times \vec{E} = -j\omega\mu\nabla \times \vec{H}.$$

There is quite a useful vector identity for a curl of a curl. In terms of the electric field,

$$\nabla \times \nabla \times \vec{E} = \nabla(\nabla \bullet \vec{E}) - \nabla^2\vec{E}.$$

It should be remembered that we guaranteed there would be no free electric charge in our medium; now, the first term on the right-hand side contains the differential form of Gauss' Law, $\nabla \bullet \vec{E} = \rho/\varepsilon$ and hence is zero. The second term represents something we did not prepare our undergraduates for in the first electromagnetics course. We solved LaPlace's equation:

$$\nabla^2 V = 0,$$

where, of course, V is a scalar. Where vectors are involved, there are three equations, one for each scalar component of the vector.

Thus, for each component of $\vec{E}$, we have

$$-\nabla^2 E_{x,y,or z} = -j\omega\mu\nabla \times H_{x,y,or z} = -j\omega\mu(j\omega\varepsilon E_{x,y,or z}).$$

It will turn out that the most important component for us will be the y-component. The wave equation for this component becomes

$$-\frac{\partial^2 E_y}{\partial x^2} - \frac{\partial^2 E_y}{\partial y^2} - \frac{\partial^2 E_y}{\partial z^2} = \omega^2\mu\varepsilon E_y.$$

Now, it is a good assumption that, with the conductivity of the walls being very high, the E-field parallel to any walls must essentially go to zero at the walls. Since there are pairs of walls at $x = 0$ and $x = a$, and also at $y = 0$ and $y = b$, the only likely conditions allowing such vanishing are that the fields go as trigonometric functions, such as

$$\sin\left(\frac{m\pi x}{a}\right) \text{ or } \sin\left(\frac{n\pi y}{b}\right),$$

where m may be any integer; the same may be said about n (not necessarily the same integer). When these functions are differentiated twice, the result is the same function multiplied by the negative coefficient of x or y squared, wiping out the first two minus signs in the equation above. Suppose we expect the z-variations to be *real* exponentials of form $e^{\pm\gamma z}$. When all these are substituted in the wave equation above and the terms grouped to see their effect, we get

$$\left(\frac{m\pi}{a}\right)^2 + \left(\frac{n\pi}{b}\right)^2 - \omega^2\mu\varepsilon = \gamma^2.$$

Now, it may become clear that we do not want a real exponential variation in the z-direction. Wave propagation requires that γ should be imaginary. This can be true for any values of m and n if the frequency is high enough that that the term overrides the other two. Each pair of values of m and n corresponds to what are called *modes*, and there is a *cutoff frequency*, which is the value of frequency where the exponent goes from real to imaginary. Most modes are of only academic interest. One mode, for which $m = 1$ and $n = 0$, has the lowest cutoff frequency of any mode; this is called the *dominant mode*, being the one most used for the transmission of energy.

Example 2.1
Find the cutoff frequency for the dominant mode in a WR 90 waveguide.

Solution: The cutoff frequency for any mode is given by the value of frequency at which the z-propagation goes from a real variation to an imaginary value. First, we may recall that the velocity of light is given in meters per second by

$$c = \frac{1}{\sqrt{\mu\varepsilon}};$$

then, the cutoff frequency is given by

$$f_{c,mn} = c\sqrt{\left(\frac{m}{2a}\right)^2 + \left(\frac{n}{2b}\right)^2}.$$

For the dominant mode, $m = 1$ and $n = 0$. Hence

$$f_{c,10} = \frac{3\times10^8\,\text{m/sec}}{2\times0.9\,\text{in}\times0.0254\,\text{m/in}} = 6.56\times10^9\,\text{Hz}.$$

Exercise 2.1
Find the cutoff frequency for the 01 and the 11 modes in a WR 90 waveguide. Remember that in WR 90, $b = 0.4$ inches.

Answers: 14.76 GHz and 16.19 GHz. Note that these answers describe the boundaries of the useful frequency range of WR 90; that is, as long as the frequency of operation is between 6.56 and 14.76 GHz, the only propagation will be of the dominant mode.

Exercise 2.2
The waveguide used at the Stanford Linear Accelerator is WR 286. Thus, $a = 2.86$ inches and $b = 1.36$ inches. Find the cutoff frequencies of the 10, 01, and 11 modes.

Answer: 2.065 GHz, 4.34 GHz, and 4.808 GHz.

2.3 WAVELENGTH OF DOMINANT MODE PROPAGATION IN WAVEGUIDE

When the frequency is above the cutoff frequency for the dominant mode, the propagation constant for waves in the z-direction is best expressed in terms of a phase shift constant given by

$$\beta = \frac{2\pi f}{c}\sqrt{1 - \left(\frac{f_c}{f}\right)^2}$$

in units of radians per meter. Now, we define wavelength as the distance a wave needs to travel to undergo a phase shift of 2π radians. We call the quantity we are seeking the *waveguide wavelength* to distinguish it from the plane wavelength, $\lambda = c/f$, and write

$$\lambda_g = \frac{2\pi}{\beta} = \frac{2\pi}{\dfrac{2\pi f}{c}} \sqrt{1 - \left(\frac{f}{f_c}\right)^2} = \frac{c}{f\sqrt{1 - \left(\dfrac{f}{f_c}\right)^2}}.$$

Exercise 2.3
Find the wavelength of the dominant wave propagation in WR 90 at 7, 9, 10, and 12.5 GHz.

Answer: 8.98 cm, 4.87 cm, 3.976 cm, and 2.819 cm.

2.4 REACTIVE CIRCUIT ELEMENTS IN WAVEGUIDES

The usual strategy is to choose the waveguide size so that at the frequency one wishes to use, only the dominant mode will *propagate*. *All* modes will be present, to a small degree; in this way, modes are analogous to higher harmonics in electronics. Fourier taught us that though we might feed pure sine waves into an amplifier, small nonlinearities in device characteristics may generate small degrees of harmonics. Similarly in waveguides, small irregularities or misalignments at joints between sections may convert small amounts of energy from the dominant mode into all the other modes, but since the higher-order modes do not propagate, the effect on the dominant mode is seen as a small reactive circuit element. During World War II, intuitive soldiers and sailors got practice in denting the waveguide with a ball-peen hammer to tune the circuit. Meanwhile, back at MIT, a number of physicists and engineers figured out how better precision could be had by closing off the waveguide with a metal vane shutting off part of the wider or narrower dimension. The results were published in *Microwave Journal* in 1966 and are shown in Figures 2.2 and 2.3.

Example 2.2: Inductive Iris
Design an inductive iris to have a normalized admittance of 1.5 at a frequency of 9.0 Mhz.

Solution: The beginner may be a bit surprised by the form of some of the parameters used in iris design, as well as in the labeling of the vertical and horizontal axes. Looking first at the vertical axis in Figure 2.2, we see the labeling is

$$\frac{B}{Y_0} \frac{a}{\lambda_g}.$$

Now, our problem statement has called out the value of B/Y_0 as 1.5. The WR 90 waveguide has $a = 0.9$ inches. So we next must find the wavelength inside this waveguide at an operating frequency of 9.0 Ghz. In Exercise 2.3 above, this was found to be 4.87 cm. Expressing all dimensions in centimeters, we may enter the vertical axis of the graph at $1.5 \times 0.9 \times 2.54/4.87 = 0.704$.

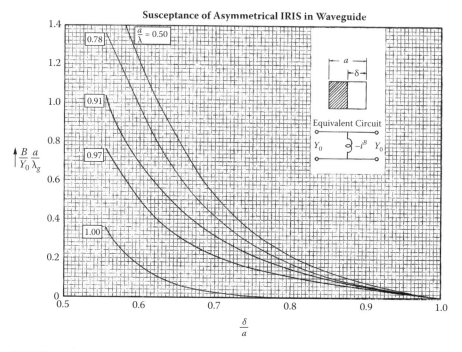

FIGURE 2.2   Inductive iris.

Next, we see there are curves drawn for several values of the parameter a/λ. Now, remember that λ is the plane wavelength, which, in vacuum, at a frequency of 9 GHz is 3.33 cm. Hence, the parameter is $a/\lambda = 0.9 \times 2.54/3.333 = 0.686$. With a very limited number of curves, suppose we make a linear interpolation between the values 0.78 and 0.5; we can say 0.686 is roughly 1/3 of the distance from 0.78 to 0.5. There are almost exactly six of the little squares to be seen. So, we read off the quantity $d/a = 0.650$ almost exactly. That ratio is the fraction of the waveguide *not* blocked off by the iris. So, now we request our machinist to make a cut into the waveguide that is

$$0.9 \text{ inch} \times (1 - 0.65) = 0.315 \text{ inch.}$$

To this we must add the wall thickness for WR 90, which is 0.050 inch, or a total cut of 0.365 inch. He or she makes the width of the cut appropriate for whatever thickness of sheet copper there is available, which perhaps another technician will silver solder into place. The whole structure then goes to the so-called plating shop, where that technician cleans it all up beautifully using some of the most toxic liquids known to man. It may then be tested to see how well we designed it.

Example 2.3: Capacitive Iris
It is certainly conceivable that one will sometimes use waveguides at power levels well below the maximum possible. In such a case, there is no compelling reason not to use a capacitive iris; so let us design one for $B/Y_0 = 1.5$, again at a frequency of 9 Ghz.

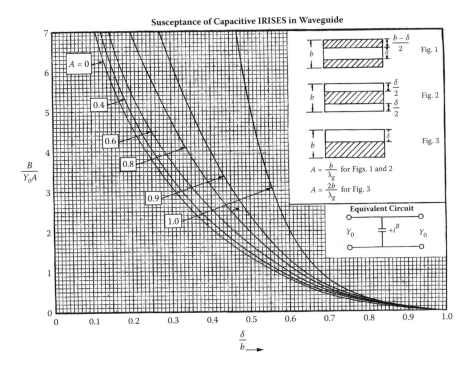

Susceptance of Capacitive IRISES in Waveguide

FIGURE 2.3   Capacitive iris.

Solution: The curves given in Figure 2.3 show three different geometries for capacitive irises, but our previous advice to keep the design as simple as possible points immediately to what is shown as "Fig. 3" on that curve. We have not changed the waveguide, frequency, or the fact that we use the dominant mode, and so the cutoff frequency and hence the guide wavelength are the same as in the other example. Thus $\lambda_{g.} = 4.870$ cm. There is only one parameter to compute,

$$A = \frac{2b}{\lambda_g} = (2 \times 0.4 \text{ inches} \times 2.54 \text{ cm/inch})/4.870 \text{ cm} = 0.417.$$

Now, we must enter Figure 2.3 on the vertical axis at $B/Y_o A = 1.5/0.417 = 3.60$. Next, the curves for $A = 0.40$ and 0.60 are very close together, so we scan across the graph for $B/Y_o A = 3.60$ to $A = 0.4$, and read the fraction of the waveguide left open as $\delta/b = 0.270$. Hence we direct our machinist to saw into the waveguide that amount plus a wall thickness (which is also 0.050 inch on that side), or

$$\text{Cut} = (1 - 0.270) \times 0.40 \text{ inch} + 0.050 \text{ inches} = 0.342 \text{ inch}.$$

Exercise 2.4

In a WR 90 waveguide, at an operating frequency of 10 Ghz, design asymmetrical irises (as we did above) each with normalized reactance $B/Y_o = 1.0$.

Partial Answers: For checking: $\lambda = 3.0$ cm, $\lambda_g = 3.9848$ cm, and $a/l = 0.339$,

$$\frac{B}{Y_0}\frac{a}{\lambda_g} = 0.256.$$

Using the the curve for $a/\lambda_g = 0.5$, one should leave 0.785 of the waveguide open, so the cut for the inductive iris should be 0.540 inch. For a capacitive iris calculation,

$$A = \frac{2 \times 0.4 \times 2.54}{3.9848} = 0.510.$$

Interpolating midway between $A = 0.4$ and 0.6, we find 0.445 of the waveguide should be open; our cut is 0.228 in.

2.5 WAVEGUIDE POWER HANDLING

Microwave devices, like those designed for lower frequencies, can be damaged if one asks them to withstand excessive voltage or current. Microwave transistors may have their highest performance at just a few percent below their highest rated current or voltage, so the engineer needs to be especially conscious of how close he/she is to the transistor ratings. Waveguides may also be susceptible to overvoltage; one may exceed the dielectric's ability to hold off voltage and arcing may occur. For dominant mode propagation, the electric field is in the y-direction and has a half-sinewave variation in the x-direction; thus

$$E_y = E_{max} \sin\left(\frac{\pi x}{a}\right),$$

and it will be accompanied by H_y, which is the same expression divided by a wave impedance:

$$Z = \frac{\sqrt{\dfrac{\mu}{\varepsilon}}}{\sqrt{1 - (f_c/f)^2}}.$$

We can obtain the transmitted power by integrating the Poynting vector over the inner cross-section of the waveguide. Therefore, in terms of the maximum electric field, the transmitted power is

$$P = \int_0^a \int_0^b \left(E_{max} \sin\left(\frac{\pi x}{a}\right)\right)^2 \sqrt{\frac{1 - \left(\dfrac{f_c}{f}\right)^2}{\mu/\varepsilon}}\, dy\, dx = (E_{max})^2 (ab)(1/4)\sqrt{\frac{1 - \left(\dfrac{f_c}{f}\right)^2}{\mu/\varepsilon}}.$$

Note that two factors of 2 resulted from integrating sine squared in time and in the x-direction. The commonly stated figure for dielectric strength for dry air at standard temperature and pressure is 3,000,000 volts/meter.

Example 2.4
The Stanford linear accelerator was designed to operate with a WR 286 waveguide, at 2856 MHz. How much power are we entitled to expect to transmit? (Note that the interior dimensions of this waveguide are 2.86 and 1.36 inches. Hence, the dominant mode cutoff is found above to be 2.08 GHz.)

Solution:

$$P = \left(\frac{3 \times 10^6}{2}\right)^2 (2.86)(1.36)(0.0254)^2 \sqrt{\frac{1 - \left(\frac{2.086}{2.856}\right)^2}{\frac{4\pi \times 10^{-7}}{8.854 \times 10^{-12}}}} = 10.22 \text{ megawatts.}$$

The beginner should be aware that this calculation assumes very idealized conditions: in particular, uniform fields are assumed, *no reflected power*, and very clean surfaces. As a taste of reality from the real world of physics, the author found that as reflections caused local power to rise above the level of *one* megawatt, one heard the characteristic sound of arcing inside the waveguide. Of course, any minor irregularity or bit of oxide or other debris on the waveguide walls could contribute to the problem and lower the power handling.

Exercise 2.5
What is the maximum power that can be transmitted through a WR 90 waveguide at 10 GHz?

Answer: 1.056 Mw.

2.6 POWER HANDLING IN COAXIAL LINES

In a coaxial line, the maximum electric field occurs at the inner conductor. In terms of the maximum value, the field between the two conductors is given by

$$E_{max}\frac{a}{r},$$

where a is the radius of the inner conductor. The voltage between the conductors is given by

$$V = \int_a^b E_{max} a \frac{dr}{r} = aE_{max} \ln\left(\frac{b}{a}\right),$$

where b is the inside radius of the outer conductor. Now, the characteristic impedance of a coaxial line is given by

$$Z_0 = \frac{60}{\sqrt{\varepsilon_r}} \ln\left(\frac{b}{a}\right),$$

so the average power flow in a wave traveling in one direction is

$$\frac{1}{2}\frac{V^2}{Z_0} = \frac{1}{2}\frac{\left[E_{max} \ln \frac{b}{a}\right]^2}{\frac{60}{\sqrt{\varepsilon_r}} \ln\left(\frac{b}{a}\right)} = \frac{\sqrt{\varepsilon_r}\left(aE_{max}\right)^2 \ln \frac{b}{a}}{120}.$$

Example 2.5

Find the maximum power that can be transmitted through an RG 58U cable. The insulation is polyethylene, which has a dielectric constant of 2.26, $a = 15$ mils (thousandths of an inch), $b = 52.5$ mils, and a handbook gives the dielectric strength of polyethylene as 460 volts/mil. So the maximum power that may be transmitted is calculated as

$$P = \frac{\sqrt{2.26}}{120}\left(460 \times 15\right)^2 \ln\left(\frac{52.5}{15}\right) = 742 \text{ kw.}$$

Don't believe it! Conditions in cables are much further from ideal than those in waveguides. If the estimate was off by a factor of ten in waveguides, there may be another factor of ten in cables.

3 Impedance-Matching Techniques

3.1 INTRODUCTION

We saw in the previous chapter that optimizing the performance of high-frequency transistors may require the connection of rather arbitrary impedances between the generator and transistor input, and between the output and the eventual 50-ohm load impedance. There are a number of techniques now available to the high-frequency circuit designer, and making a choice may depend upon whether the frequency of operation is VHF, UHF, or microwave; how broad a band of frequencies must be handled by the design; the availability of chip components of appropriate physical size and circuit value; and other problems. When microwave circuits were just coming into use, from World War II onward, the main method was first to compute the distance from the load at which a purely shunt reactive circuit element would perfectly match the transmission line. Then the reactance was fabricated as a *stub*, which was another length of transmission line with a good solid short as its load, or in waveguides, as a kind of window in which a thin slice of metal closed off part of the waveguide. In both of these cases, the reactance shunted the line, that being the only physically feasible connection for the coaxial cable and the waveguide. More recently, the widespread use of microstrip lines, in which the signal-bearing conductor is very accessible, has made it very reasonable to connect a *surface mount* capacitor or inductor in *series* with the line, so more flexibility is available to the circuit designer. We will start with a technique that should be entirely understandable to the EE (electrical engineering) student as soon as AC circuit analysis has become a useful part of his/her analysis tools.

3.1.1 MATCHING IMPEDANCES USING REACTIVE L-SECTIONS

Let us hasten to state that by an *L-section* we mean not only inductances, but that the circuit diagram will have reactive elements drawn at right angles to each other, as in the capital letter "L." Figure 3.1 shows two basic connections for these two "L's." In each of the cases diagrammed above, we have a resistive load that we wish to modify using the matching section to match to a different real impedance. We will find that the first diagram has the effect of matching the load to a larger impedance. The analysis and design are straightforward and quick.

3.1.2 MATCHING LOWER IMPEDANCE TO HIGHER IMPEDANCE

Refer to part (a) of Figure 3.1 We write the admittance of X, in series with R, as

$$Y = \frac{1}{R_L + jX_1} \times \frac{R_L - jX_1}{R_L - jX_1} = \frac{R_L}{R_L^2 + X_1^2} - \frac{jX_1}{R_L^2 + X_1^2}.$$

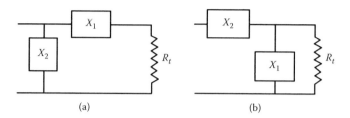

(a)   (b)

FIGURE 3.1   Two possible orientations of L-sections to achieve impedance match. (a) Real load impedance less than the resistance to be matched. (b) Real load impedance greater than the resistance to be matched.

The magnitude of X_1 is thus chosen to bring the real part of this admittance to the design value. Then X_2 is chosen of the right magnitude *and sign* to reduce the total imaginary part of the admittance to zero. Note that to produce the right magnitude of the real part of admittance, the sign of X_1 matters not at all, because it is squared in the expression for the real part of admittance. However, whatever the sign of X_1, X_2 must be chosen of the correct magnitude *and* sign to cancel the imaginary part of the admittance of the series connection.

Example 3.1

A 20-ohm load is to be modified by the L-section to present a real part of 50 ohms impedance, at 200 MHz, using a series capacitor and a shunt inductor. Find the magnitude of each reactive element.

Solution: We need

$$\frac{R_L}{R_L^2 + X_1^2} = \frac{1}{50},$$

when $R_L = 20$ ohms, so

$$\frac{20}{400 + X_1^2} = \frac{1}{50}.$$

Cross-multiplying, $400 + X_1^2 = 20 \times 50 = 1000$. Therefore,

$$X_1^2 = 600; \; X_1 = 24.49 = \frac{1}{\omega C} = \frac{1}{4\pi C \times 10^8}; \text{ and solving for } C, C = 32.5 \text{ pF.}$$

The magnitude of reactive admittance to be matched is

$$\frac{24.49}{(400 + 600)} = \frac{1}{\omega L_2}; \; L_2 = \frac{1000}{24.49 \times 4 \times 10^8 \pi} = 32.5 \text{ nH.}$$

Exercise 3.1

Find the magnitudes of the matching elements if X_1 is inductive.

Answer: $L_1 = 19.5$ nH and $C_2 = 19.5$ pF.

Exercise 3.2

Find the magnitude of the elements that will match a 50-ohm load to 75 ohms, at a frequency of 500 MHz

Answer: 11.25 nH in series and 3.00 pF in parallel or 9.00 pF in series and 33.76 nH in parallel with the series combination.

3.1.3 MATCHING A REAL LOAD THAT HAS GREATER RESISTANCE THAN SOURCE

With knowledge of the behavior of duals, we might expect that an L-section that first shunts the load impedance will reduce the effective real part of impedance, and fortunately it does work that way. Referring now to part (b) of Figure 3.1, the impedance of the parallel combination is

$$\frac{jX_1 R_L}{R_L + jX_1} \times \frac{R_L - jX_1}{R_L + jX_1} = \frac{X_1^2 R_L + jX_1 R_L^2}{R_L^2 + X_1^2} = \frac{R_L}{1 + \left(\dfrac{R_L}{X_1}\right)^2} + \frac{jX_1}{1 + \left(\dfrac{X_1}{R_L}\right)^2}.$$

Thus we shunt the load with reactance until the real part of impedance is correct, after which we add the series reactance of the right sign to reduce the total reactance to zero.

Example 3.2

Find the value of shunt capacitance and series inductance to match 80 ohms real to a 50-ohm source at 500 MHz.

Solution: We need

$$\frac{80}{1 + \left(\dfrac{80}{X_1}\right)^2} = 50; \; 50\left(\frac{80}{X_1}\right)^2 = 30; \; X_1 = 80\sqrt{\frac{5}{3}} = 103.3;$$

$$\frac{103.3}{1 + \frac{5}{3}} = 10^9 \pi L; \; L = 12.33 \text{ nH.}$$

Exercise 3.3

Find the shunt inductance and series capacitance that will make the match of impedance needed in the example above.

Answers: 32.87 nH and 8.22 pF.

Exercise 3.4

Design two possible L-sections to match a load of 100 to 75 ohms.

Answers: 55.1 nH shunt and 7.35 pF series, or 1.84 pF shunt and 13.78 nH series.

3.2 USE OF THE SMITH CHART IN LUMPED IMPEDANCE MATCHING

When the Smith chart is augmented to show lines of both constant normalized impedance and admittance simultaneously, it can be very helpful in understanding the mechanics of lumped-circuit impedance matching. In general, one should normalize all impedances to the generator impedance to which they are to be matched, if that impedance is purely real. Consider the examples in the previous section. Normalizing the 20-ohm load to 50 ohms gives $r = (20/50) = 0.4$; so we enter the Smith chart at point A on Figure 3.2. We may move in *either direction* along the $r = 0.4$ line until we reach the $g = 1.0$ line at either B or B'. From the chart, it appears that we had to add a normalized reactance of precisely 0.5. It is almost always likely to be preferable to make the more precise numerical calculation, which recommends adding a normalized reactance of $x = \pm 0.49$. Since the frequency is 200 MHz, the series element is either

$$L = \frac{.49 \times 50}{4\pi \times 10^8} = 19.5 \text{ mH, or}$$

$$C = \frac{1}{4\pi \times 10^8 \times 0.49 \times 50} = 32.48 \text{ pF.}$$

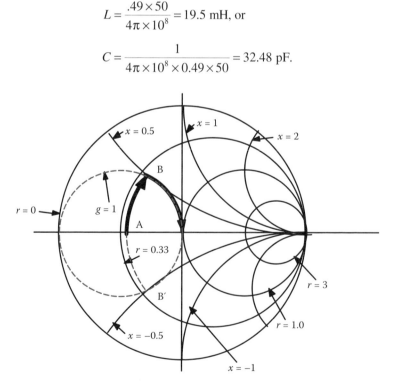

FIGURE 3.2   Transforming real impedance upward using series reactance.

At the location where $z = 0.4 \pm j0.49$, we can now calculate the normalized admittance as $y = 1 \pm j1.225$, a number that would be borne out by careful scrutiny of the two-color Smith chart. Thus, if our shunt element were an inductor, the match would be completed using a series capacitor given by

$$1.225 = 50 \ \omega C; \ C = \frac{1.225}{50 \times 4\pi \times 10^8} = 19.5 \text{ pF.}$$

If the shunt element was a capacitor, the impedance match could be completed using a series inductor figured as follows:

$$1.225 = \frac{50}{\omega L}; \ L = \frac{50}{1.225 \times 4\pi \times 10^8} = 32.48 \text{ nH.}$$

Next, let us consider the case where the load to be matched is 80 ohms; we enter the Smith chart, noting, as seen in Figure 3.3, where $z = 1.6$, hence $y = g = 0.625$. We may move either upward or downward along the .625 line, although since it is not drawn on the graph, it appears that the imaginary part of the admittance, called susceptance and given the symbol "b," is approximately ± 0.5 at the point, corresponding to 100 ohms. Of course, the more accurate numerical calculation would yield 103.3 ohms. As we saw before, to obtain the impedance of 103.3 ohms, we required 3.08 pF or 32.87 nH at 500 MHz.

Again, from the Smith chart, we may read the normalized impedance after we add the shunt element as $z = 1 \pm j.75$; this would mean we must add ±37.5 ohms reactive to cancel the reactance we have, but, again, the more accurate numerical

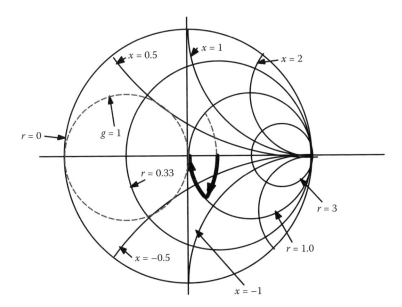

FIGURE 3.3   Transforming real impedance downward using shunt reactance.

calculation would call out 38.7 ohms. Clearly, the accuracy of the graphical calculation is limited, but in view of the available magnitudes of the components one can
buy, the graphical accuracy may be sufficient.

Perhaps the most important advantage of using the Smith chart on these
impedance-matching problems is that it helps the engineer to visualize a surprising number of additional possibilities, which might not otherwise have occurred
to him/her. First, we should note that $z = 20 \pm j$, and anything may in principle be
matched by adding a series inductor or capacitor of the correct amount to make the
normalized reactance $= \pm.49$. Having done that, we put in parallel the same inductor
or capacitor that was needed in the original design. In examples where we start with
some small load resistance, practical considerations may suggest that the series reactance one adds should be the choice that requires the smaller amount of reactance;
that is, if the load reactance is small and inductive, one should add more inductance
to reach the desired impedance.

When the normalized admittance or impedance to be matched lies outside both
the $r = 1$ and $g = 1$ circles, an interesting new possibility is found—the number of
matching possibilities doubles once again. To stay on somewhat familiar ground, let
us consider for the moment impedances with the real part of 20 ohms; the added possibilities occur when $x > .49$ (of course, when $x > .49$ we can equally well start with
a shunt resistance). Let us consider, for example, a load with a normalized impedance of $z = r + jx = 0.4 - j0.8$. (Please refer to Figure 3.4.) Certainly, we could again
add $x = .31$ to arrive at $0.4 - j.49$ as before. Then, also as before, the normalized
admittance is $y = 1.0 + j1.225$, so that adding more inductance to give an admittance
of $-j1.225$, matches the line. However, we could also start by first changing admittance to get the real part of normalized *impedance* to unity. The original normalized
admittance is $0.5 + j1.0$. We could add a normalized admittance of $-j0.5$, arriving at

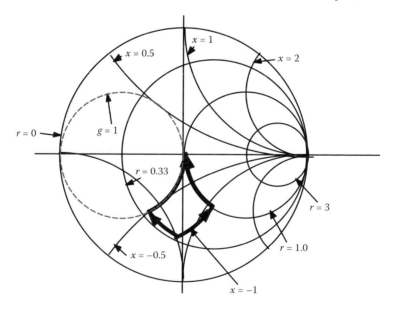

FIGURE 3.4   Illustrating two trajectories by which $z = 0.4 - j0.8$ may be matched.

$y = 0.5(1 + j)$, which corresponds to a normalized impedance of $1 - j$. Again we must add inductance, this time in *series*, to cancel the negative reactance. The magnitude of these elements at 500 MHz is

$$\text{shunt L} \quad 0.5 = \frac{Y}{Y_o} = \frac{Z_o}{\omega L},$$

$$L = \frac{Z_o}{0.5\omega} = \frac{50}{0.5 \times 10^9 \pi} = 31.8 \text{ nH},$$

$$\text{series L} \quad 1.0 = \frac{\omega L}{Z_o},$$

$$L = \frac{Z_o}{\omega} = 15.9 \text{ nH}.$$

Theoretically, of course, one could add considerably more shunt inductance to arrive at the point

$$y = 0.5 - j0.5 \text{ and } z = 1 + j1.$$

However, the more drastic change in the circuit is probably much less practical; the larger the change in the circuit for the purpose of impedance matching, the faster the situation deteriorates as the frequency changes. Of course, in this latest solution, with this shunt element, the matching would be completed by adding a series *capacitor*.

Exercise 3.5
Find the L-section elements to match a load of $10 + j30$ ohms to a 50-ohm source in two ways. Frequency is 800 MHz.

Answer: Series C = 19.89 pF, shunt C = 7.96 pF or shunt capacitance of 3.98 pF followed by series C of 3.98pF.

Exercise 3.6
Find two L-sections that will match a load of $15 + j45$ ohms to a 50-ohm line at 800 MHz.

Answer: 9.04 pF in series with 6.09 pF parallel with the combination, or 2.1 pF in parallel, followed by 2.8 pF in series with the combination.

3.3 IMPEDANCE MATCHING WITH A SINGLE REACTIVE ELEMENT

Because impedance varies as one moves along a transmission line, one can match the line using a single reactive element if one finds the correct location. One simply moves along the line to a point where the real part of impedance is the desired value;

it is then possible to work with series-matching elements. If it is intended that a *shunt* reactive element be used to match the line, it is necessary to go to the position on the line where the real part of *admittance* is already the correct value.

Let us illustrate what is needed in an example. Let us see what is required to match a 100-ohm resistor to a 50-ohm line at 500 Mhz. It is our plan to find the location nearest the load where adding a single parallel reactance will match the line; therefore, we first find the nearest location to the load terminals where normalized admittance $y = 1.0 \pm jb$. We first normalize the load impedance and enter the Smith chart at that point. The normalized load impedance is

$$z_t = \frac{Z_t}{Z_0} = \frac{100}{50} = 2.0.$$

We use the Smith chart in Figure 3.5 to graph the variation of normalized imped- ance as one moves from the load terminals toward the generator. $z = 2.0$ is on the positive real axis. Making the usually good assumption that attenuation effects can be neglected in a line a fraction of a wavelength long, we rotate clockwise on the circle $|G(d)| = 1/3$ until we first intersect the $g = 1.0$ line at approximately $y = 1 + j0.7$. The line will be matched from the latter location to the generator if we shunt the line with $y = -j0.7$ at this location, clearly an inductor, given by

$$0.7 = \frac{Y}{Y_0} = \frac{Z_0}{\omega L}; \ L = \frac{50}{0.7 \times 10^9 \pi} = 22.7 \text{nH}.$$

If we work as carefully as possible with a full-size Smith chart, we find that we have moved from 0.250 to 0.417 wavelengths toward the generator, or 0.167 λ. A careful

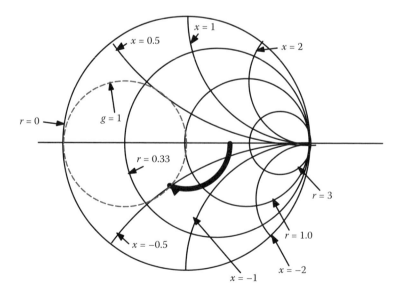

FIGURE 3.5   Moving on Smith chart from $r = 2.0$ to $g = 1 + jb$.

computation on a \$100 calculator gives $b = 0.7071$ and a distance from load to matching element of exactly one-sixth of a wavelength. We enter the chart on the negative real axis at $g = 0.5$. We rotate on a circle of constant radius toward the generator, until we encounter the circle where $y = 1 \pm 0.71$; this requires moving a distance of $0.152\ \lambda$. At that location, the line is matched if we add a shunt reactance having a normalized admittance of -0.71. Calculating the amount of inductance required,

$$\frac{y}{y_o} = Y;\ z_o = .71 = \frac{50}{\omega L};\ L = \frac{50}{.71\omega} = \frac{50}{.71 \times 10^9 \pi} = 22.4\ \text{nH}.$$

Knowing the characteristics of the transmission line we are matching, we can figure the location of the matching reactance in centimeters, inches, or whatever is appropriate. Almost all coaxial lines that one could encounter are insulated with stabilized polyethylene, for which the dielectric constant is 2.26. Hence, the velocity of waves on the line at high frequency is given as

$$\frac{c}{\sqrt{2.26}} = 1.996 \times 10^8\ \text{m/sec.}$$

Thus, the distance to the matching reactance, previously found on the Smith chart to be $0.152\ \lambda$, can now be calculated as

$$d_{match} = .152 \times \frac{1.996 \times 10^8}{5 \times 10^8} = 6.1\ \text{cm.}$$

If the transmission line is a microstrip, one must determine the effective dielectric constant, as was demonstrated in Chapter 1. One then computes the wavelength using this effective value.

Exercise 3.7
Find the location and magnitude of a matching reactive element to match 150 ohms to a 50-ohm coaxial line at 600 MHz.

Answer: 3.78 cm from load, 11.5 nH.

Exercise 3.8
Find the location and magnitude of matching reactance to match 15 ohms to a 75-ohm microstrip line for which $\varepsilon_r = 2.0$. The frequency is 750 MHz.

Answer: 1.87 cm from load, 5.0 pF.

Exercise 3.9
Find the location and magnitude of a reactive element to match $50 - j100$ ohms to a 50-ohm line. Frequency is 600 MHz, and the line is a microstrip for which the effective dielectric constant is 3.5.

Answer: 6.63 nH located 3.37 cm from the load.

Exercise 3.10

Find the location and magnitude of the reactive element to match $25 + j25$ ohms to a 50-ohm coaxial line. The frequency is 750 MHz.

Answer: 4.2 pF located right at the load.

3.4   STUB MATCHING

At one time reactive elements suitable for matching microwave transmission lines did not exist. The technique used was called *stub matching*. The required reactance was provided by a short length of transmission line having a short or open circuit as its termination. With a reflection coefficient having a magnitude of unity, the input impedance of this short length of transmission line, which is called a *stub*, is purely reactive. Hence, the appropriate magnitude of reactance is made by the appropriate choice of its length.

In principle, either a shorted or open-circuited transmission line would serve equally well as an impedance-matching section; in practice, it may be much easier to fabricate a good solid short than an open one. Too often, a signal-bearing conductor that is not grounded may act as a fairly efficient antenna. If energy is lost by radiation, a *radiation resistance* will be seen, which means that the input impedance will not be purely imaginary.

The technique of stub matching may be illustrated using the same example in the previous section. There we found that in matching 100 ohms to a 50-ohm line, we first found the location, 0.152λ, from the load. At that point, the normalized input impedance was $y = 1 \pm j0.71$. The stub must thus provide $b = -.71$.

Let us return to theory to consider the input impedance of a short-circuited lossless transmission line. For a short circuit termination,

$$G_t = \frac{0 - z_o}{0 + z_o} = -1.$$

Then,

$$Z(d) = Z_o \frac{1 + G_t e^{-j2\beta d}}{1 - G_t e^{-j2\beta d}} = Z_o \frac{e^{j\beta d} - e^{-j\beta d}}{e^{j\beta d} + e^{-j\beta d}}.$$

The sum and difference of positive and negative exponentials are conveniently expressed in terms of trigonometric functions:

$$Z(d) = Z_o \frac{j2\sin\beta d}{2\cos\beta d} = j Z_o \tan\beta d.$$

Since we usually require some normalized admittance, we write

$$y(d) = \frac{1}{\dfrac{Z(d)}{Z_o}} = \frac{1}{j\tan\beta d} = \frac{-j}{\tan\beta d}.$$

In the example, we require

$$\frac{-j}{\tan \beta d} = -j0.71,$$

$$\beta d = \frac{2\pi d}{\lambda} = \tan^{-1} \frac{1}{.71} - \tan^{-1} 1.414 = .957 \text{ rad},$$

$$d = \frac{.957\lambda}{2\pi} = .152\lambda.$$

The reader may be very impressed by the fact that the stub length and distance from the load came out to be the same in this example. *This is sheer coincidence* and will not happen again. Young engineers should not expect to see any appreciable amount of real magic in their careers.

Exercise 3.11
Find the lengths of stubs made with 50-ohm insulated line with polyethylene, which would replace the discrete reactances in the exercises of the previous section.

Answers:

Stub susceptance of $b = -1.1547$ is provided by a length of 3.78 cm.
Stub susceptance of $b = 1.7889$ is provided by a length of 11.85 cm.
Stub susceptance of $b = -2$ is provided by a length of 2.46 cm.
Stub susceptance of $b = 1$ is provided by a length of 9.98 cm.

3.5 MATCHING SECTIONS

3.5.1 QUARTER-WAVELENGTH MATCHING SECTIONS

There is one more traditional method of matching loads to transmission lines or to generators that, in its simplest form, matched a real impedance to another real impedance, and therefore might not be considered versatile enough for general use. One example of its success is shown in Figure 3.6. On this Smith chart, we see an arc rotating one-half of the way around the chart, which has the effect of transforming a normalized impedance of 3.0 to one of 1/3.0. It can be seen that it is necessary to choose the characteristic impedance of the quarter-wavelength section appropriately. Given two real impedances that one wishes to match to each other, it is not difficult to prove that the characteristic impedance of the quarter-wavelength section needs to be the geometric mean of the two impedances. One example of how this could work would be to use a load impedance of 225 ohms and a generator impedance of 25 ohms. The characteristic impedance of the matching section would then need to be

$$Z_0 = \sqrt{Z_1 Z_2} = \sqrt{225 \times 25} = 75 \text{ ohms}.$$

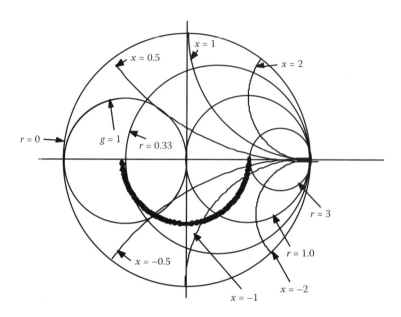

FIGURE 3.6   Quarter-wavelength line transforming $r = 3$ to $r = 1/3$.

Exercise 3.12

Find the impedance of the quarter-wavelength matching sections to match a load of 150 ohms to a generator impedance of 50 ohms, a 200-ohm load to a 50-ohm generator, and a 750-ohm load to a 120-ohm generator.

Answers: 86.6 ohms, 100 ohms, and 300 ohms.

3.5.2 MATCHING WITH OTHER THAN QUARTER-WAVELENGTH

For many years this writer thought that he and his students were the only people in the world who had noticed that a complex impedance can be matched to a real one by using matching sections of more or less than one quarter-wavelength. Finally, an author named Rizzi* succinctly specified that a complex load impedance $R + jX$ can be matched to a generator or transmission line having a source or characteristic impedance Z_0 by using a *matching section* having characteristic impedance

$$Z_{01} = \sqrt{\frac{Z_0\left(R(Z_0 - R) - X^2\right)}{Z_0 - R}}.$$

Rizzi also gives the length of the matching section as one given by

$$\tan \beta l = Z_{01}\frac{Z_0 - R}{XZ_0}.$$

---

* Peter A. Rizzi, *Microwave Engineering, Passive Circuits* (Englewood Cliffs, NJ: Prentice Hall, 1988), 132–133.

There are one or two possible complications that may arise in the use of these formulas. In the characteristic impedance formula, the minus signs present may lead to a maximum value of reactance that can be matched; greater values of reactance lead to a purely imaginary characteristic impedance, which is of course not available. In the formula for the length of the matching section, first, the dimensions of βd *must* be radians. Another unfortunate characteristic of calculators and most computer programs is that if either is asked to find arc tan of something negative, it will give an answer in the fourth quadrant as a negative angle. Such an answer would imply a negative length of matching section, which makes no physical sense. The reader is urged to obtain an answer in the *second* quadrant in such cases; the simplest way may be to add pi radians to the answer in the fourth quadrant.

Example 3.3
Let us check out this method. A rather bulky antenna called the *rhombic* has an input impedance of $800 + j175$. Find the characteristic impedance and the length of the matching section to match it to 300 Ω. (Note: 300 Ω is the characteristic impedance of the balanced two-wire line called the *twinlead*.)

Solution: Substituting in the expression above,

$$Z'_o = \sqrt{\frac{300(800^2 - 300 \times 800 + 175^2)}{800 - 300}} = 508.3 \ \Omega.$$

Next, let us find the length of the matching section in a way that capitalizes on our knowledge of the Smith chart. First we normalize the load impedance by dividing it by the characteristic impedance *of the matching section* and plotting it on the Smith chart. We then rotate on an arc of constant reflection coefficient to the negative part of the real axis, obtaining $l = 0.283\lambda$ as the length of the matching section (see Figure 3.7). We know that we must end up on either the positive or negative real axis, where the input impedance is purely real; since the amount of the normalized impedance is 0.590, we know we want the *negative* real axis. The point we plotted is found to be 0.283 wavelengths toward the load on the Smith chart, which is then the length of the matching section in wavelengths. Let us check this result using Rizzi's formula. We have

$$\tan \beta d = \frac{508.3 \times (300 - 800)}{175 \times 300} = -4.841.$$

For such a tangent, the HP 48S gives $\beta d = -1.3671$. Since we cannot use this answer, we add pi, obtaining $\beta d = 1.7745$. Now we must use the relation between the wavelength and the phase shift constant, $\beta = 2\pi/\lambda$, and solve for the length d of the matching section, $l = 0.2824 \ \lambda$, which agrees with our graphical determination on the Smith chart decently well.

Exercise 3.13
Match $200 - j50$ to a 300-ohm line.

Answer: $Z_{01} = 230 \ \Omega$, length of 0.342 λ.

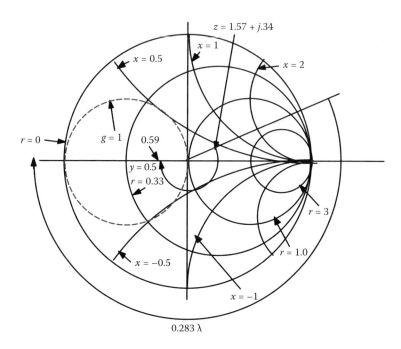

FIGURE 3.7   Smith chart normalized to 508 ohms for matching load of $800 + j175$ ohms to 300 ohms.

Exercise 3.14

Find the characteristic impedance and length of a matching section to match a load impedance of $35 + j7$ to a 50-ohm line. Note that while load resistance is less than the impedance to be matched, the reactance here is not large enough to make the match impossible in principle.

Answer: 39.8 Ω, 0.166 λ.

Exercise 3.15

Repeat Exercise 3.13 for a load of $100 + j70$.

Answer: 150.5 Ω, 0.152 λ.

3.6   WHEN AND HOW TO UNMATCH LINES

In Chapter 1 we found that load and source impedances needed to be far from 50 ohms for some transistors to be conjugate matched at both the input and output ports. That makes it sound a little as though one may be in the business of changing a nice real 50 ohms into something complex, and one may wonder if he/she knows how to do that.

As is usual on transmission lines, we may obtain $Z(d)$ as

$$Z(d) = Z_0 \frac{1 + \Gamma(d)}{1 - \Gamma(d)}.$$

Hence, if one sets up for a simultaneous impedance match at the input and the output, the source impedance "seen" by the transistor will be

$$Z_S = Z_0 \frac{1 + \Gamma_{MG}}{1 - \Gamma_{MG}} = 50 \frac{1 + 0.890\angle - 178.71°}{1 - 0.890\angle - 178,71°} = 2.91 - j0.561.$$

Of course, the implication of this result is that the input impedance of the transistor will be $2.91 + j0.561$ if Γ_{ML} is connected as the load. Hence, the design goal of the engineer is to modify a real impedance of 50 ohms so as to obtain $2.91 - j0.561$. His or her procedure is first to observe that the real part of impedance must be greatly reduced. If the engineer decides to accomplish this using an L-section, the first step is to connect a reactance across the 50-ohm input impedance. As was found in the early part of this chapter, the amount of the reactance is given by

$$2.91 = \frac{50}{1 + (50 / X_1)^2}.$$

Solving this equation for X_1, one obtains the *magnitude* of the reactance then as

$$|X_1| = 12.43 \text{ ohms.}$$

We have not absolutely determined the *sign* of the reactance. The other relation determined at the beginning of the chapter was

$$X_2 = \frac{X_1}{1 + (X_1 / 50)^2}.$$

The first thing to be noted here is that the sign of X_2 will be the same as for X_1. From this circuit, we need a final reactance that is negative, so the first principle here is that every time we make a choice, it should get as close as possible to the final result. Certainly, we could use a positive reactance and cancel it out with a much larger negative reactance, but better engineering practice would be to start with the correct sign and simply add to it. From the equation just above, if we start with $-j12.43$ in parallel with 50 real ohms, the result is $X_2 = -11.71$. But we needed only a total of -0.56, so we must add $+j11.15$ ohms in series to get the correct result.

We have proven that the transistor can be provided the impedance it desires. It might be comforting to prove that the generator, which has a purely real impedance of 50 ohms, "sees" the input impedance it expects. It may be helpful to refer to

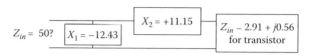

FIGURE 3.8   Matching L-section at the input.

Figure 3.8. The following reinforces our statements concerning what is connected where:

We have connected $j11.15$ in series with $2.91 + j0.56$, obtaining a total of $2.91 + j11.71$.

Now we wish to know the *admittance* of this series combination, so we take the reciprocal of this series impedance, obtaining $Y = 0.01999 - j0.0804$. This is in parallel with a reactance $Z = -j12.43$, which corresponds to an admittance of $j0.08045$, so the reactances cancel each other and we are left with a real impedance that corresponds very closely to 50 ohms. (It should be noted that the small discrepancies one finds here result from pulling numbers off the calculator and substituting them with limited accuracy.)

Exercise 3.16
Repeat the example above at the output of the transistor, where one is looking for $\Gamma_{ML} = 0.777\angle66.10°$.

Answers: One needs $Z_L = 20.34 + j72.92$ ohms. Because this real part is smaller than 50 ohms, one first needs to parallel the 50 ohms with the appropriate reactance, as was done in Section 3.1.3; this would be $|X_1| = 41.41$. The series equivalent of this parallel combination is $20.34 + j24.56$, if the sign of X_1 was chosen as positive. One still must add series reactance to arrive at the right total; $X_2 = 48.36$ would be the right amount to add.

Exercise 3.17
The values of Γ_{ML} and Γ_{MG} used above were found for the MRF 571 transistor at 1 Ghz. Use this knowledge to find the kinds and magnitudes of lumped circuit elements to perform the conjugate impedance matching.

Answers: At the input, parallel the generator with capacitance of 12.8 pF. Then place an inductor of 1.77 nH in series with the parallel combination. At the output, parallel the 50-ohm load with $j41.41$, which will require an inductor of 6.59 nH. In series with this, we need a reactance of $j48.36$, which calls for an inductor of 7.7 nH.

Exercise 3.18
Find the lengths and characteristic impedances of matching sections to produce the correct driving and load impedances, that is, $Z_G = 2.91 - j0.56$ and $Z_L = 20.34 + j72.92$ ohms.

Answers: At input, $Z_{01} = 12.08$ Ω, length of 0.2578 λ. At output, the matching section method won't work because the reactance is too high.

Exercise 3.19
Assume the matching section is to be made using microstrip material for which the dielectric constant is 2.25. If the thickness of the dielectric is 1 millimeter, find the width of the trace for Z_{01} found in Exercise 3.18, the effective dielectric constant, and the length of the matching section at $f = 1$ GHz.

Answers: $W = 1.82$ cm, $\varepsilon_{eff} = 2.11$, and length = 5.32 cm.

Exercise 3.20
Find the distance from the input or the load terminals of the above transistor where a single series reactance may be placed in a 50-ohm line to conjugate-match input and output lines. Also, find the kind and the magnitude of the reactance. Assume that the line is a microstrip as analyzed in Section 1.7, having $\varepsilon_{eff} = 1.905$.

Answers: 0.2105λ or 4.57 cm for input terminals. We need $x = -3.904$, or $-j195.2 \, \Omega$, which could be provided by 0.815 pF. Go 0.354λ from the load terminals or 7.69 cm, and add $x = -2.47$, or $-j123.4 \, \lambda$, which can be provided by 1.29 pF.

Exercise 3.21
Find the distance from the input or the load terminals of the above transistor where a single short-circuited shunt stub in a 50-ohm line may be placed in the 50-ohm line to conjugate-match input and output lines. Also, find the physical length of the stub.

Answers: In the input line, go 0.0359λ or 0.78 cm from the input terminals. For a normalized input admittance of $j3.904$, we need 0.230λ or 5.00 cm. In output line, go 0.104λ or 2.26 cm. For a normalized stub admittance of $y = -j2.47$, use a stub length 0.0306λ or 0.67 cm.

3.7   CIRCUIT TERMINALS FOR WAVEGUIDES

When one is mounting components in waveguide, it may be somewhat unclear where the load terminals are. A certain amount of ingenuity may be required to know where to place impedance-matching elements. The configuration of waveguide circuits is usually oneport or twoport style in which a flange is welded to each open port so that assemblies may be made by bolting them together. Thus, it is most convenient to specify the location of matching elements in relation to the location of the surface of the flanges.

In the early days of waveguides, there did not exist the powerful vector network analyzers that had been developed by the end of the 20th century. What was used to measure normalized impedance were "slotted lines," which were a length of coaxial line or waveguide mounted on a structure having a carriage that could be moved along the line. Also on the carriage was an antenna extending into a longitudinal slot in the coaxial line or waveguide. Sensing the electric field in the line gave an indication of total voltage, which is at maximum values when the incident and reflected waves are in phase and minimum values when they are 180 degrees out of phase. Figure 3.9 shows the variation of total voltage on a line where unity represents

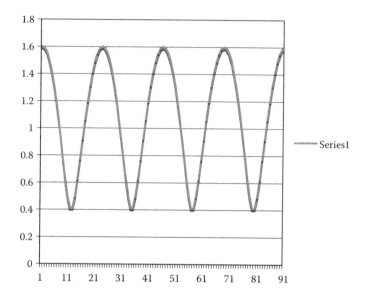

FIGURE 3.9   Voltage standing wave pattern, VSWR = 4.0.

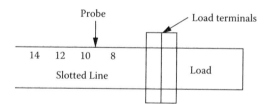

FIGURE 3.10   Detail of slotted line near load terminals.

amplitude of the incident wave; the pattern shows a standing wave pattern where the voltage standing wave ratio (usually abbreviated VSWR) of 4.0 to 1. It should be noted that the pattern repeats itself with impressive regularity, so the impedance may be matched in relation to some location where it is known.

  The standard technique for waveguide impedance match was, indeed still may be, to first attach a polished brass plate to the terminals of the slotted line and determine the locations of standing wave minima; since the standing wave pattern repeats every <u>half</u> wavelength the wavelength is just twice the distance between adjacent minima. Let us assume, for example, that when this was done VSWR was very high (effectively infinite); with wonderful luck, we found the locations of minima at 10.0, 12.0 and 14.0 cm. Suppose that the pattern of Figure 3.9 was obtained with an unknown load connected, but minima were at 10.4, 12.4 and 14.4 cm. Note that the calibration of the typical slotted line has the numbers increasing as one moves away from the load terminals, as in Figure 3.10.

Since the normalized impedance is the reciprocal of the VSWR at a VSWR minimum, it is real and equal to ¼ = 0.250 at 10.4 cm on the slotted line. We need to know it at 10.0, so we need to go a distance of

$$\frac{10.4 - 10.0}{4} = 0.1\lambda_g$$

toward the load on the Smith chart. We enter the Smith chart at r = 0.25, x = 0 and rotate on a circle of constant radius 0.1 wavelength towards the load. We obtain z = 0.37 + j0.67.

Exercise 3.19
In the example above, find the location nearest the flange where an inductive iris may be located and what its susceptance would need to be to match the load.

Answers: Note that irises always <u>shunt</u> the line. Hence, we must find the location where normalized admittance is y = 1 + j(positive number) The first interesting location on the Smith chart would be at 0.176 "wavelengths toward load," but note that at the location noted, normalized <u>impedance</u> is 1 + j1.5. <u>Admittance</u> will have that value one quarter wavelength away at 0.426 wavelengths "toward the load." Since the flange location was at 0.1 wavelengths, we subtract 0.1, and end up at 0.326 wavelengths, or 1.304 centimeters from the flange.

Exercise 3.20
The author received his early experience in microwave circuits designing for the frequency of the Stanford Linear Accelerator. The frequency of design was 2856 MHz. The medium of the design was WG 286, for which the inside dimensions are a = 2.86 inches and b = 1.36 inches. Wall thicknesses are 0.070 inches for all walls. Find the dimensions of an asymmetric inductive and capacitive iris (note that this means figure "c" in Figure 2.3) in this waveguide, for this frequency and for normalized admittances of y = 2.0

Answers, intermediate and final: Cutoff frequency for dominant mode is 2.065 GHz. Waveguide wavelength is 15.20 cm.
In the inductive iris design, the parameter

$$\frac{B}{Y_0}\frac{a}{\lambda_g} = 0.955.$$

The parameter $a/\lambda = 0.69$. The fraction of the waveguide left open is $\delta/a = 0.62$. Hence, the depth of the cut should be 0.948 inch.
In the capacitive iris design, the parameter

$$A = \frac{2b}{\lambda_g} = 0.455.$$

$$\frac{B}{AY_0} - 2.20.$$

The fraction of the waveguide left open is $\delta/b = 0.41$. Hence, the depth of the machinist's cut is 0.872 inch.

3.8 LOCATION OF IRIS FROM STANDING WAVE MEASUREMENTS

Refer to Figure 3.12, on which the numbers on the slotted line increase as one moves away from the load terminals; they may thus be considered to register d, which we have identified as the distance of the point from the load terminals. The first part of any standing wave measurement might be to put a good short circuit on the load terminals and determine the location of the minima on the slotted line. Let us say they are at 8, 10, and 12 cm.

Example 3.6
Now, suppose we connect our previously unmeasured load and find VSWR = 2.0, with minima at 9.5, 11.5, and 13.5 cm. Where should we locate an iris and what should be its susceptance?

Solution: First, we know that the distance between the adjacent minima on the slotted line is *one-half* wavelength, thus $\lambda_g = 2 \times (10-8)\,\text{cm} = 4\,\text{cm}$. For this VSWR, the magnitude of the reflection coefficient is

$$\left| \Gamma_t \right| = \frac{2-1}{2+1} = \frac{1}{3}.$$

So we enter the Smith chart at $\Gamma(d) = 0.333\angle 180°$ or at what is the same thing, $z = 0.5 + j0$. In terms of the markings on the Smith chart, we can say we are at 9.5, 11.5, or 13.5, whichever is convenient for us. If we need to know the load impedance referred to at the load terminals, we must move to 8.0 or 10.0 cm or the like. From 9.5 to 10 is 0.5 cm, or 0.125_{-g} *toward the generator*. When we do this, we find $z_L = 0.8 + j0.6$. Now, suppose we say we go from the load terminals to the *first location in the load structure where a shunt susceptance of either sign will match the line.* It turns out that there may be an untenable difficulty with this first location on the $g = 0$ circle. We find that the input admittance at this location is $y = 1 - j0.71$. From the load terminals, we have gone only $0.027\,\lambda_0 = 1.08$ mm toward the load. However, there is a husky flange by which one bolts the load to a similar flange on the slotted section, both of which are approximately 4.5 mm thick. Clearly, if you ask the technician to saw into the flange, he will mention to your boss that he hired a real dud this time, and your boss will suspect that you were missing on the day they did this calculation in your college lab. With superior wisdom, you say you will go where $y = 1 + j0.71$, for which you must travel $0.222\,\lambda_g$ or 8.88 mm. Of course, to again avoid the scorn of the machine shop, be sure to convert those millimeters to inches, and tell the man to make the cut 0.351 inch from the terminal side of the flange. From the iris you need $y = -j0.71$, which is inductive. Now that you know how, it is left as an exercise to figure out the depth of the cut. For convenience, assume the operating frequency is 10 Ghz.

Partial Answers: Enter Figure 3.10 on the vertical axis at 0.180, read off (doing a small interpolation) 0.805 when the waveguide is open. Therefore, cut $0.195 \times 0.9 + 0.050$ inch or 0.226 inch.

4 Scattering Coefficients of Two-Ports

4.1 HYBRID TWO-PORT PARAMETERS

The reader may (we hope) remember the two-port parameters that were introduced at the end of their network analysis course. Two-ports have also been called *four-terminal devices*. Two each of the terminals are associated with one or the other of the ports, often called the input or the output ports. The *passive circuit assumption* is usually used to define the positive or negative voltage or current at each port, as illustrated in Figure 4.1. We have assigned signs such that positive currents flow into the positive voltage terminal at each port. A number of sets of two-port parameters have been defined; they may be impedances, admittances, or a mixture. First we will illustrate one of the latter, since low-frequency transistor specifications are usually given using *hybrid* or *h-parameters*. The equations defining these parameters are

$$v_1 = h_{11}i_1 + h_{12}v_2$$

$$i_2 = h_{21}i_1 + h_{22}v_2.$$

The dimensions of these parameters may be inferred when one reasons how to derive or measure them. In the equations, suppose we make $v_2 = 0$. Then

$$h_{11} = v_1/i_1, \text{ an impedance}$$

$$h_{21} = i_2/i_1 \text{ is a current gain.}$$

If one makes $I_1 = 0$, then h_{12} is a voltage gain and h_{22} is an admittance.

Let us calculate the h-parameters for three impedances arranged in a tee as shown in Figure 4.2. In the equation discussion above, we first proposed making $v_2 = 0$. In the circuit to be analyzed, we would make $v_2 = 0$ by short-circuiting terminal 2. We would then have the circuit connection shown in Figure 4.3. In this circuit, h_{11} would be the input impedance resulting from Z_A in series with the parallel combination of Z_B and Z_c, or

$$h_{11} = Z_A + \frac{Z_B Z_C}{Z_B + Z_C}.$$

Next, we have a current division; I_1 divides between Z_B and Z_C. The current division theorem says that the fraction of the total current through one of two parallel

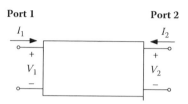

FIGURE 4.1 Two-port network showing positive voltages and currents.

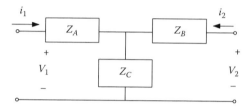

FIGURE 4.2 Two-port comprising a "tee" connection of impedances.

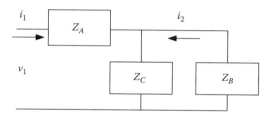

FIGURE 4.3 T-circuit with the two terminals shorted.

impedances is given by the *other* impedance over the sum, and the beginner might look at the circuit and say

$$i_2 = i_1 \frac{Z_C}{Z_B + Z_C}.$$

The only thing wrong with this relation is its algebraic sign; if I_1 is positive, a positive I_2 would be the current flowing *upward* through Z_B, so it is in fact negative. Hence

$$h_{21} = -\frac{Z_C}{Z_B + Z_C}.$$

To determine the other two parameters, we need to make $I_1 = 0$. To assure that $I_1 = 0$, we leave the number 1 terminals open, and we connect *nothing* to the number 1 terminals. The pertinent variables are seen below, where we have simply reviewed Figure 4.2, noting that $I_1 = 0$. Leaving terminals 1 open means that no current flows in Z_A, hence there is no voltage drop in Z_A. Therefore, V_1 is the voltage across Z_C and we have a *voltage* division between Z_B and Z_C. Therefore,

$$h_{12} = \frac{Z_c}{Z_b + Z_c}.$$

h_{22} is the *admittance* seen between terminals 2, so

$$h_{22} = \frac{1}{Z_B + Z_C}.$$

Exercise 4.1

Find the h-parameters for the "pi" connection of the three impedances shown in Figure 4.4.

Answers: $h_{11} = \dfrac{Z_A Z_C}{Z_A + Z_C}$; $h_{21} = -\dfrac{Z_A}{Z_A + Z_C} = -h_{12}$; $h_{22} = \dfrac{1}{Z_B} + \dfrac{1}{Z_A + Z_C}$.

Exercise 4.2

Find the h-parameters for the simplified low-frequency equivalent circuits shown in Figure 4.5 for a bipolar transistor.

Answer: $h_{11} = r_{b'e} + r_{bb''}$, $h_{21} = g_m r_{b'e}$, $h_{12} = 0$, $h_{22} = 1/r_o$.

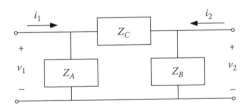

FIGURE 4.4 Two-port comprising three impedances in a pi connection.

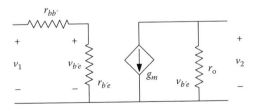

FIGURE 4.5 Simplified low-frequency equivalent circuit for bipolar transistor.

4.2 NOW, ABOUT THOSE SCATTERING PARAMETERS ...

The early engineers trying to get high-frequency performance out of bipolar transistors found that transistors did not respond well to having their terminals shorted *or* left open. In either case, the transistors might become unstable, which is to say they might oscillate, producing an output when there is no input. This is not a completely unacceptable behavior; oscillators are what one uses to obtain test signals. In about 1939, two unemployed Stanford grads named Bill Hewlett and Dave Packard set to work in a garage to make an audio oscillator for Walt Disney to use in the movie *Fantasia*. Recently the computer manufacturing part of the company became so large that they split off the part making test equipment, called it Agilent, and are still growing.

Those early engineers reasoned that any kind of load could be drastically changed by the impedance transformations produced on transmission lines. The best way to bring stability to testing transistors was found to be terminating the transmission lines connected to transistors in their characteristic impedances. The equivalent circuit taking into account the wave behavior on the lines connected to a transistor is as shown in Figure 4.6.

The variables labeled a and b are, respectively, the voltage amplitudes of waves directed at and away from the two-port. The equation set describing the relationships between the parameters is

$$b_1 = S_{11}a_1 + S_{12}a_2$$

$$b_2 = S_{21}a_1 + S_{22}a_2.$$

Now, for example, if we make $a_2 = 0$, then we can get S_{11} as the input reflection coefficient at terminals 11 through 1. Since a_2 is the voltage of a wave traveling toward the transistor in transmission line 2, we can make $a_2 = 0$ by connecting a load of Z_o to those terminals.

Example 4.1

Let us illustrate some of the necessary reasoning by first figuring the scattering coefficients for a very simple case, then for increasing complexity. Let our first sample two-port have the normally grounded terminals connected together, with a resistor between the other ports. We will also make $a_2 = 0$ by connecting Z_0 to those

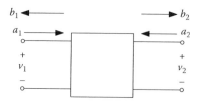

FIGURE 4.6 Waves directed at and away from a high-frequency transistor.

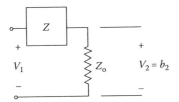

FIGURE 4.7   Impedance Z from input to output, terminals 2 terminated with Z_0.

terminals as shown in Figure 4.7. We can now write S_{11} as the input reflection coefficient at terminals 1. From transmission line theory, we can write the reflection coefficient in terms of the input impedance, as

$$S_{11} = \Gamma_{in} = \frac{Z_{in} - Z_0}{Z_{in} + Z_0} = \frac{Z + Z_0 - Z_0}{Z + Z_0 + Z_0} = \frac{Z}{Z + 2Z_0}.$$

The transmission coefficient is

$$S_{21} = \frac{b_2}{a_1} = \frac{V_1 \times \dfrac{Z_0}{Z + Z_0}}{a_1} = \frac{a_1 \times (1 + \Gamma_{in}) \dfrac{Z_0}{Z + Z_0}}{a_1}$$

$$= \left(1 + \frac{Z}{Z + 2Z_0}\right)\left(\frac{Z_0}{Z + Z_0}\right) = \frac{Z + 2Z_0 + Z}{Z + 2Z_0} \times \left(\frac{Z_0}{Z + Z_0}\right)$$

$$= \frac{2(Z + Z_0)}{Z + 2Z_0} \times \left(\frac{Z_0}{Z + Z_0}\right) = \frac{2Z_0}{Z + Z_0}.$$

If we now connect Z_0 to terminals 1 and apply voltage to terminals 2, our circuit would be a mirror image of what we just analyzed. Thus, we may say for this circuit that $S_{11} = S_{22}$ and $S_{12} = S_{21}$.

Example 4.2
Next, let us find the scattering coefficients of a bipolar transistor for which the simplified low-frequency equivalent circuit is as shown in Figure 4.5. Assume the standard characteristic impedance of 50 ohms.

Solution: To obtain S_{11}, we solve the circuit assuming a 50-ohm load connected to the output. It can be seen that there is no mechanism here for feeding back any voltage from output to input, that is, the input circuit is independent of anything happening at the output. Hence, the input impedance is $(25 + 200)$ ohms $= 225$ ohms, and the input reflection coefficient, which will also be S_{11}, is

$$\Gamma_{in} = S_{11} = \frac{225 - 50}{225 + 50} = \frac{175}{225} = \frac{7}{11} = 0.636.$$

At the output, we have 50 ohms in parallel with 4000 ohms, yielding about 49.38 ohms. The output voltage is then $-0.25 \times 49.38\, v_{b'e} = -12.35\, v_{b'e}$, and is also equal to b_2 because we terminated the number 2 terminals with Z_0, making the reflected wave, which is a_2, equal to 0. Now, in terms of the input voltage, $v_{b'e} = (200/225)\, v_1 = (8/9)\, v_1$,

$$v_1 = (1 + \Gamma_{in})\, a_1 = (1 + 7/11)\, a_1 = (18/11)\, a_1.$$

So, we have $b_2 = -12.35\, v_{b'e} = -12.35 \times (8/9)v_1 = -12.35 \times (8/9)(18/11)a_1 = -17.96a_1$.

Hence, $S_{21} = 17.96\angle 180°$. If now we connect nothing but 50 ohms to the input, this causes no $v_{b'e}$; therefore, the dependent current source is zero, with the result that $S_{12} = 0$. The reflection coefficient looking "back" into the output is

$$S_{22} = \frac{4000 - 50}{4050} = 0.9753\angle 0°.$$

Exercise 4.3
Find the scattering coefficients for a simple shunt impedance as shown in Figure 4.8.

Answer: $S_{11} = S_{22} = -Z/(2Z + Z_0)$ and $S_{21} = S_{12} = 2Z/(2Z + Z_0)$.

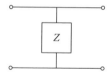

FIGURE 4.8   Simple shunt resistor.

Exercise 4.4
Now add the capacitances of the hybrid-pi high-frequency circuit to the low-frequency circuit previously analyzed, and evaluate the scattering coefficients at 1, 10, 100, and 1000 Mhz. Note that we have arranged for a beta cutoff frequency around 100 Mhz, and that the capacitance from output to input will make for a nonzero S_{12} (see Figure 4.9).

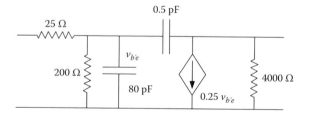

FIGURE 4.9   High-frequency equivalent circuit for high-frequency transistor.

Example 4.3 (at 10 Mhz)
We will first solve a circuits problem assuming 1 volt connected to the left-hand (or the number 1) terminals and another 50-ohm load connected to the right-hand (or number 2) terminals. If the bottom rail is the reference node, we have just two unknown node voltages, say $v_{b'e}$ and v_2. We simply sum the currents flowing out of each of these nodes, thus:

$$\frac{v_{b'e}-1}{25} + \frac{v_{b'e}}{200} + j1.6\pi \times 10^{-3} v_{b'e} + j10^{-5}\pi(v_{b'e} - v_2) = 0,$$

$$\frac{v_2}{4000} + \frac{v_2}{50} + 0.25 v_{b'e} + j10^{-5}\pi(v_2 - v_{b'e}) = 0.$$

We will multiply the first equation by 200, move the one volt to the right-hand side, and multiply the second equation by 4000, getting

$$v_{b'e}(8 + 1 + j(0.32\pi + 0.002\pi)) - j0.002\pi \, v_2 = 8$$

$$v_{b'e}(1000 - j0.04\pi) + v_2(81 + j0.04\pi) = 0.$$

We might find it preferable to have the polar form of complex numbers. Then the equations would be

$$v_{b'e}(9.0567\angle 6.413°) + V_2(.0628\angle 90°) = 8$$

$$v_{b'e}(1000\angle - 0.0072°) + v_2(81\angle 0.0889°) = 0.$$

Solving the simultaneous equations by any convenient method, the student may compare his/her answers to the author's results of

$$v_{b'e} = 0.889\angle - 1.501° \text{ and } v_2 = 10.97\angle 178.338°.$$

Of course, we did all this to determine the scattering coefficients at this frequency. S_{11} is the input reflection coefficient,

$$\Gamma_{in} = \frac{Z_{in} - Z_0}{Z_{in} + Z_0}.$$

Since the input voltage was one volt, we can get Z_{in} as $1/i_{in}$, where

$$i_{in} = \frac{1 - v_{b'e}}{25} = 0.00455\angle 11.80°.$$

Then $Z_{in} = 219.64\angle - 11.80°$ and

$$S_{11} = \frac{219.64\angle - 11.80° - 50\angle 0°}{219.64\angle - 11.80° + 50\angle 0°} = 0.636\angle - 5.60°.$$

Now, $S_{21} = b_2/a_1$, where $b_2 = v_2$, because we made $a_2 = 0$ by connecting Z_0, although the one volt we assumed contains both a_1 and b_1. However, we can say $1 = a_1 + b_1 = a_1(1 + S_{11})$ and thus

$$a_1 = \frac{1}{1 + S_{11}} \quad \text{and } S_{21} = v_2(1 + S_{11}) = 17.94\angle 176.16°.$$

These magnitudes are very close to the DC values quoted above, but the capacitors have begun to cause a bit of a phase shift.

Now, we recall that

$$S_{12} = \frac{b_1}{a_2},$$

so what we need to do is work our way through some voltage dividers. By inspection, we can say that

$$b_1 = \frac{50}{75} v_{b'e}$$

in which $v_{b'e}$ is itself a division of v_2 appearing across a complex parallel impedance. The 75 Ω in parallel with 200 Ω gives an equivalent R in parallel with the 80 pF, of 54.54. The impedance of R in parallel with C can be massaged to a less hairy form of $R/(1 + j\omega RC)$, and then

$$v_{b'e} = \frac{\dfrac{R}{1 + j\omega RC}}{\dfrac{R}{1 + j\omega RC} + \dfrac{1}{j\omega C_{b'c}}} v_2 = \frac{R}{R\left(1 + \dfrac{C}{C_{b'c}}\right) + \dfrac{1}{j\omega C_{b'c}}} v_2; \quad v_1 = \frac{2}{3} v_{b'e}.$$

Hence

$$S_{12} = \frac{0.66667 \times 54.54}{161 \times 54.54 - \dfrac{j10^5}{\pi}} = 0.001101\angle 74.576°.$$

Next we set out to write the output admittance by writing i_2 in terms of v_2;

$$i_2 = \frac{v_2}{4000} + v_{b'e}(0.25) + (v_2 - v_{b'e}) j\omega C_{b'c}$$

$$= v_2\left(\frac{1}{4000} + j\omega C_{b'c}\right) + v_{b'e}(0.25 - j\omega C_{b'c}).$$

However, remembering that $v_1 = (2/3) \, v_{b'e}$ and also $v_1 = S_{12}v_2$, we may write

$$v_{b'e} = 1.5 \, S_{12}v_2,$$

so we may write the output admittance by writing i_2 in terms of v_2:

$$\frac{1}{4000} + j\omega C_{b'c}' + 1.5S_{12}(0.25 - j\omega C_{b'c}') = 7.98 \times 10^{-4} \angle 63.23°.$$

The resulting reflection coefficient will be

$$S_{22} = \frac{1/Y - 50}{1/Y + 50} = 0.965 \angle - 4.08°.$$

Exercise 4.5
Find S-parameters at 100 Mhz and 1 Ghz.

Answers: 100 Mhz: $S_{11} = 0.388 \angle -121.736°$, $S_{21} = 6.564 \angle 110.408°$, $S_{12} = 0.00388 \angle 19.92°$, $S_{22} = 0.850 \angle -4.66°$. 1 Ghz: $S_{11} = 0.3345 \angle -173.48°$, $S_{21} = 0.695 \angle -98.07$, $S_{12} = 0.00414 \angle 2.07°$, $S_{22} = 0.879 \angle -21.19°$.

4.3 TWO-PORT HIGH-FREQUENCY MODELS FOR BIPOLAR TRANSISTORS

The HP8753 Network Analyzer will graph each scattering parameter over the range of frequency chosen. If one needs the parameters at certain specific frequencies, one can put the cursor on each frequency desired and the magnitude and phase of the complex number will be read out. Motorola provides data in both formats; some data copied from *Motorola RF Device Data** (a sort of house organ not in general circulation) are also graphed in the complex plane, as shown in Figure 4.10.

Example 4.4
From the scattering coefficients, find the input and output impedances when the other terminals are terminated in a 50-ohm load, for the MRF571 at 200 Mhz.

Solution: We can consider $S_{11} = 0.74 \angle - 86°$ and $S_{22} = 0.69 \angle - 42°$ to be the load reflection coefficients at one end of the transistor when the other end is loaded by 50 ohms real. Hence,

$$Z_{in} = Z_0 \frac{1 + S_{11}}{1 - S_{11}} = 50 \frac{1 + .74 \angle - 86°}{1 - .74 \angle - 86°} = 15.6 - j51.11 \text{ ohms, and}$$

$$Z_{out} = Z_0 \frac{1 + S_{22}}{1 - S_{22}} = 50 \frac{1 + .69 \angle - 42°}{1 - .69 \angle - 42°} = 50.14 - j102.47 \text{ ohms.}$$

---

* Motorola, Inc. *Motorola RF Device Data Vols. I and II*, 1988: Author.

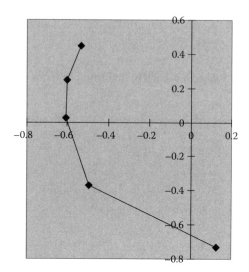

FIGURE 4.10   A plot of parameter S_{11} at several frequencies.

4.4   INPUT AND OUTPUT IMPEDANCES: GENERAL CASE

Just above, it was slightly artificial to obtain input and output impedances by terminating the terminals not being fed by the generator with Z_0. As we look at the defining equations for S-parameters, this made $a_2 = 0$, so that the input reflection coefficient became S_{11}. Still looking at the equation for b_1, we see that we would also obtain

$$\Gamma_{in} = S_{11} \text{ if } S_{12} \to 0.$$

This would mean that the transistor was called *unilateral*, that is, signal flows through it in only one direction. If one compares the values for the scattering coefficients at various frequencies, it is noticeable that S_{12} increases as the frequency increases. This should not be too surprising, as it arises from capacitance, which in the bipolar transistor is $C_{b'c}$ and, of course, capacitive admittance increases with frequency, allowing more and more signal to be fed from the output circuit back to the input. However, one may require an expression for the input reflection coefficient that is valid for all conditions. Suppose we relate a_2 and b_2 through a reflection coefficient, to wit,

$$a_2 = \Gamma_{out}b_2.$$

We substitute thus for b_2 in the second defining equation:

$$\frac{a_2}{\Gamma_L} = S_{21}a_1 + S_{22}a_2.$$

This needs to be solved for a_2 and the result substituted in the other defining equation:

$$a_2 \left(\frac{1}{\Gamma_L} - S_{22} \right) = \left(\frac{1 - S_{22}\Gamma_L}{\Gamma_L} \right) a_2 = S_{21}a_1; \quad a_2 = \frac{S_{21}\Gamma_L \, a_1}{1 - S_{22}\Gamma_L}.$$

Now, substituting in the first defining equation,

$$b_1 = S_{11}a_1 + \frac{S_{12}S_{21}\Gamma_L a_1}{1 - \Gamma_L S_{22}}.$$

Next, defining the input reflection coefficient,

$$\Gamma_{in} = \frac{b_1}{a_1} = S_{11} + \frac{S_{21}S_{12}\Gamma_L}{1 - \Gamma_L S_{22}}.$$

One can obtain an absolutely analogous expression for Γ_{out}, postulating that the input terminals will be terminated with a reflection coefficient,

$$\Gamma_G = a_1/b_1,$$

obtaining, through similar complex algebra,

$$\Gamma_{out} = S_{22} + \frac{S_{21}S_{12}\Gamma_G}{1 - S_{11}\Gamma_G}.$$

If one further examines the scattering coefficients as a frequency increases, one sees that the amplifying coefficient S_{21} approaches unity or less. Thus, the gain of a stage could be small indeed, and care must be taken to match impedances as much as possible. Indeed, the best one could possibly do would be to have both input and output circuits simultaneously conjugate matched, which is to say, at both input and output of the transistor, the input impedance is the complex conjugate of the generator impedance, and the output impedance is the complex conjugate of the transistor output impedance. The author has not been able to track down the unrecognized engineer who first slogged through a jungle of complex algebra to derive the conditions for a simultaneous conjugate match. If the reader wishes to expose him/herself to this derivation before taking advantage of it, Pozar* gives a fairly thorough modern treatment. It should be noted at the outset that it is only possible to have input and output simultaneously conjugate matched if the transistor is *unconditionally* stable at that frequency. Pozar and others provide us with the necessary conditions for unconditional stability. First, the determinant of the scattering matrix must have a magnitude less than unity. Defining the determinant as Δ, then $\Delta = S_{11}S_{22} - S_{21}S_{12}$. Note that nowhere in this expression have we written a simple magnitude, but the first stability criterion is $|\Delta| < 1$. The other necessary stability criterion is $|K| > 1$, where K is defined as

$$K = \frac{1 - |S_{11}|^2 - |S_{22}|^2 + |\Delta|^2}{2|S_{12}S_{21}|}.$$

---

* David M. Pozar, *Microwave Engineering*, 2nd ed. (New York: John Wiley & Sons, 1998), chap. 11.

If both these conditions are met, one can calculate the values of Γ_{out} and Γ_{in} such that there will be simultaneous conjugate matches at both sets of terminals. There are some intermediate results that make the formulas at least look more approachable. They are

$$B_1 = 1 + |S_{11}|^2 - |S_{22}|^2 - |\Delta|^2$$
$$B_2 = 1 - |S_{11}|^2 + |S_{22}|^2 - |\Delta|^2$$
$$C_1 = S_{11} - \Delta S_{22}^*$$
$$C_2 = S_{22} - \Delta S_{11}^*.$$

Then, the load reflection coefficient for conjugate matching at both sets of terminals is

$$\Gamma_{LM} = \frac{B_2 - \sqrt{B_2^2 - 4|C_2|^2}}{2C_2}.$$

The comparable reflection coefficient at the input end is

$$\Gamma_{GM} = \frac{B_1 - \sqrt{B_1^2 - 4|C_1|^2}}{2C_1}.$$

The transducer gain we may expect if we meet all these conditions is then given by

$$G_{TM} = \left(|S_{21}|/(|S_{12}|)\right) \times \left(K - \sqrt{K^2 - 1}\right).$$

Example 4.5
Let us now compute all these numbers for the MRF 571 transistor, at $I_C = 5.0$ mA, $V_{CE} = 6.0$v, and at a frequency of 1 Ghz.

Solutions: For the conditions stated, we can read

$$S_{11} = 0.61\angle178°$$
$$S_{21} = 3.0\angle78°$$
$$S_{12} = 0.09\angle37°$$
$$S_{22} = 0.28\angle-69°.$$

Hence, $\Delta = (0.61\angle178°)(0.28\angle-69°) - (3.0\angle79°)(0.09\angle37°) = 0.1017\angle-54.89°$.
  Then

$$K = \frac{1 - (0.61)^2 - (0.28)^2 + (0.1017)^2}{2(3.0)(0.09)} = 1.0367.$$

Thus we have established that the transistor *is* unconditionally stable at this frequency, so we may proceed to find the reflection coefficients for a simultaneous conjugate match:

$B_1 = 1 + (0.61)^2 - (0.28)^2 - (0.1017)^2 = 1.2833$

$B_2 = 1 + (0.29)^2 - (0.61)^2 - (0.1017)^2 = 0.7017$

$C_1 = 0.61\angle 178° - (0.1017\angle - 54.89°)(0.28\angle + 69°) = 0..6374\angle 178.71°$

$C_2 = 0.28\angle - 69° - (0.1017\angle - 54.89)(0.61\angle - 178°) = 0.3400\angle - 66.10°.$

At the input, the magnitude of the reflection coefficient for simultaneous match is

$$|\Gamma_{GM}| = \frac{1.2783 - \sqrt{1.2783^2 - 4(.6374)^2}}{2 \times 0.6374} = 0.890.$$

Now, since C_1 appears in the denominator as the only complex number in the equation, the phase of Γ_{GM} is the negative of the phase of C_1; therefore,

$$\Gamma_{GM} = 0.890\angle - 178.71°$$

$$|\Gamma_{LM}| = \frac{0.7017 - \sqrt{(0.7017)^2 - 4(0.34)^2}}{2 \times (0.34)} = 0.777.$$

Its phase is of course the negative of the phase of C_2, or 66.10°. The gain that can be expected if these reflection coefficients are used is

$$G_{TM} = \frac{3}{0.09}\left(1.0367 - \sqrt{(1.0367)^2 - 1}\right) = 25.44, \text{ representing 14.06 dB.}$$

Exercise 4.6
Demonstrate that conjugate match has been obtained in the example above, that is, let $\Gamma_L = \Gamma_{LM}$ and calculate Γ_{in}, hoping to get the conjugate of Γ_{GM}. Also, let $\Gamma_G = \Gamma_{GM}$ and calculate Γ_{out}, hoping to get the conjugate of Γ_{ML}.

Answers: They come out really close to the correct results. Repeat either or both until you get them right.

Exercise 4.7
Perform all the calculations of the example above for the MRF 571 transistor at 1500 Mhz.

Answers:

$$\Delta = 0.1296\angle - 23.97°, \ K = 1.197.$$

Hence, they are unconditionally stable at this frequency. Therefore,

$\Gamma_{MG} = 0.803\angle - 161°$

$\Gamma_{ML} = 0.6075\angle 70.45°$

$G_{TM} = 9.8 \rightarrow 9.9 \text{ dB.}$

4.5 STABILITY CIRCLES FOR THE POTENTIALLY UNSTABLE TRANSISTOR

The author has forgotten which old Woody Allen movie it was where he and the lady were prowling around a scary environment when Woody says, "This is no time for panic." Five seconds later they find a scary thing and he says, "*Now* is the time for panic!" as they begin their flight. Similarly, the microwave engineer might panic prematurely when he/she finds frequencies where a transistor is not unconditionally stable. Fortunately, geometric constructions called *stability circles*, when drawn on a Smith chart, are useful for showing what values of input and/or output reflection coefficients must be avoided to keep the transistor stable. The circle that applies at the transistor input* is centered at

$$C_G = \frac{(S_{11} - \Delta S_{22}^*)^*}{|S_{11}|^2 - |\Delta|^2}$$

and has a radius given by

$$R_G = \left| \frac{S_{12}S_{21}}{|S_{11}|^2 - |\Delta|^2} \right|,$$

where this writer writes G for the generator end rather than Pozar's S for "sending." At the load end, we have

$$C_L = \frac{(S_{22} - \Delta S_{11}^*)^*}{|S_{22}|^2 - |\Delta|^2} \text{ and } R_L = \left| \frac{S_{12}S_{21}}{|S_{22}|^2 - |\Delta|^2} \right|.$$

It turns out that the MRF 571 is potentially unstable at 500 Mhz. Let us find the warning phenomena, and then determine the stability circles to find regions of safety in the following example.

Example 4.6

Compute Δ and K for the MRF 571 transistor at 500 Mhz to find the potential instability. Then calculate the center and radius of each stability circle and sketch them on a Smith chart.

Solution: At 500 Mhz, we have

$S_{11} = 0.62\angle-143°$
$S_{21} = 5.5\angle97°$
$S_{12} = 0.08\angle33°$
$S_{22} = 0.41\angle-59°$
$\Delta = (0.62\angle-143°) \times (0.41\angle-59°) - (5.5\angle97°) \times (0.08\angle33°) = 0.2464\angle-78.97°$

---

* Pozar, *Microwave Engineering*, 2nd ed. (New York: John Wiley & Sons, 1998), chap. 11.

Since $|\Delta| < 1$, one of the necessary conditions for unconditional stability is met:

$$K = \frac{1+(0.2464)^2 - (0.62)^2 - (0.41)^2}{2 \times 5.5 \times 0.08} = 0.5775.$$

This result is not greater than one, so the transistor fails one of the necessary conditions for unconditional stability. However, instead of simply throwing up our hands, we draw stability circles:

$$C_G = \frac{(0.62\angle -143° - (0.2464\angle -78.97°)(0.41\angle 59°))^*}{(0.62)^2 - (0.2464)^2}$$

$$= 2.102\angle 150.15° = -1.82 + j1.05$$

$$R_G = \frac{5.5 \times 0.08}{(.62)^2 - (.2464)^2} = 1.36$$

$$C_L = \frac{(0.41\angle -59° - (0.2464\angle -78.97°)(0.62\angle 143°))^*}{(0.41)^2 - (0.2464)^2}$$

$$= 4.75\angle 73.55° = 1.34 + j4.55$$

$$R_L = \frac{5.5 \times 0.08}{(0.41)^2 - (0.2464)^2} = 4.1.$$

Parts of the stability circles are sketched in Figure 4.11, where the origin is the center of the Smith chart.

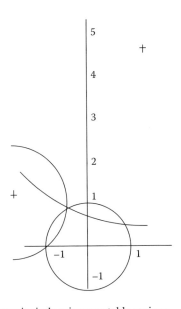

FIGURE 4.11 Smith chart unit circle minus unstable regions.

Exercise 4.8

Repeat the exercise above at 200 Mhz, where

$S_{11} = 0.74\angle 86\,°$
$S_{21} = 10.5\angle 129°$
$S_{12} = 0.06\angle 46°$
$S_{22} = 0.69\angle -42°.$

Answers: $\Delta = 0.5543\angle -55.89°$, $K = 0.225$ (potentially unstable); $C_G = 1.33 + j2.70$, $R_G = 2.62$, $C_L = 0.94 + j3.96$, $R_L = 3.73$

Exercise 4.9

The consequence of using a reflection coefficient in the "forbidden" areas defined by the stability circles is that input or output reflection coefficients are greater than one, meaning more energy comes out than goes in. From the example above, check the input reflection coefficient for $\Gamma_L = 0.8\angle 70°$, and the output reflection coefficient for $\Gamma_G = 0.9\angle 150°$.

Answers: $\Gamma_{in} = 1.128\angle -148.46°$, $\Gamma_{out} = 1.281\angle -67.41°$.

5 Selective Circuits and Oscillators

Most readers have probably been exposed to selective circuits in the tuned circuits section of an AC circuits course. However, such a treatment is often rushed and rather light on practical aspects, especially high-frequency phenomena. Of course, unless his/her company is in the business of producing selective circuits, not too many engineers will be required to do any tuned-circuit design. However, it is worthwhile to introduce some of the vocabulary and ways of specifying the performance of selective circuits. Also, while the engineer needing a receiver may be able to buy integrated circuits containing electronics and fixed-tuned circuits, he/she may still need to design for a particular input frequency and against specific possible interferences. Thus, this chapter will go from the performance of rather simple single-tuned circuits to the rather high performance of piezoelectric crystal filters. In the latter part of the chapter, resonance will be employed as the main determinant of the operation frequency of several types of oscillators.

5.1 LRC SERIES RESONANCE

Series resonance is analyzed here, not because it is very often put to use, but because later equivalent circuits are most easily understood in terms of the series resonator. Consider the circuit in Figure 5.1. We can write the transfer function relating output voltage over input voltage by simply applying voltage division, saying that the desired ratio is the capacitive impedance over the total series impedance:

$$\frac{V_o}{V_i}(j\omega) = \frac{1/j\omega C}{j\omega L + R + 1/j\omega C} = \frac{1}{1 - \omega^2 LC + j\omega CR}.$$

Since the denominator contains a real and an imaginary part, we might suspect that the maximum of the transfer function would occur at the value of ω for which the real part goes to zero. This value, at which the inductive and capacitive impedances exactly cancel each other out, is called the *frequency* (in radians per second) *of series resonance*. We can find that the transfer function can peak up rather sharply if the value of resistance is small compared to the value of either reactance.

Example 5.1

Let us plot the magnitude of the transfer function versus frequency for $L = 1$ H, $C = 1$ μF and for several rather different values of resistance, to wit, 20, 50, and 1000 ohms.

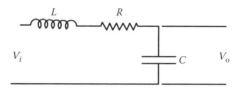

FIGURE 5.1   Series resonant circuit.

Solution: Perhaps the most striking aspect of the graphs in Figure 5.2 is that for the lowest value of R, the ratio of output to input voltage is a maximum of 50. For that or perhaps other reasons, this quantity was originally called the *quality factor*, abbreviated *Q*, of the circuit. Because the source of *R* is in fact mainly the unavoidable resistance of the wire from which the inductor is wound, the *Q* for series resonance is usually defined as

$$Q = w_0\,L/R.$$

Perhaps even more significantly, it is related to the half-power bandwidth of the circuit. Clearly, the transfer function drops to 0.707 of its peak value when the real part of the denominator is equal to the imaginary part. Using also the expression for resonant frequency, we may write the equation for the transfer function as

$$H(j\omega) = \frac{1}{1 - \omega^2/\omega_0^2 + j\omega/\omega_0 Q}.$$

Setting the real part equal to the imaginary part, we obtain the two positive values for half-power frequency as

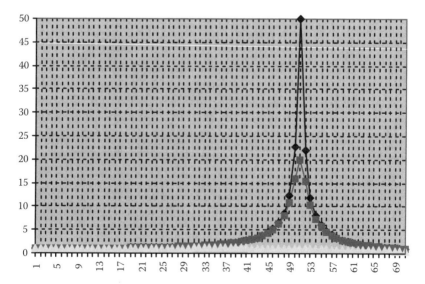

FIGURE 5.2   Series resonator capacitor voltage for several values of *R*.

$$\omega_{2,1} = \sqrt{\omega_0{}^2 + \left(\frac{\omega_0}{2Q}\right)^2} \pm \frac{\omega_0}{2Q}.$$

One can obtain the half-power radian bandwidth, which this author likes to call W, by subtracting the lower half-power frequency from the higher. Since both expressions contain the radical, they subtract, and we are left with

$$W = \omega_0/2Q - (-\omega_0/2Q) = \omega_0/Q.$$

This may be stated as follows: The fractional half-power bandwidth in a resonant circuit is given by $1/Q$.

Example 5.2
Suppose the 10-kHz bandwidth of an intermediate frequency (IF), centered at 455 kHz, for an AM receiver is determined by one tuned circuit. What must be the Q of the tuned circuit?

Solution: We can get the required Q by simply dividing the center frequency by the bandwidth. Hence, $Q = 455$ kHz/10 kHz = 45.5.

Exercise 5.1
Repeat the example above for RF amplifiers at each end of the AM broadcast band, that is, at center frequencies of 530 and 1650 kHz.

Answers: 53 and 165. It is fortunate that one will seldom need to have the latter Q, as it may be far from available.

Exercise 5.2
Repeat the example for bandwidths of 200 kHz in FM IF amplifiers (centered at 10.7 Mhz) and the high end of the band (107.9 Mhz).

Answers: 53.5 and 539.5. (IF filters are often prefabricated using the high-Q capabilities of so-called ceramic filters, and one would have to be excessively innocent to attempt the latter filter.)

Exercise 5.3
If we are aiming to pass the entire assigned 6 Mhz, what Q would be required for channel 2 centered at 57 Mhz, channel 13 at 213 Mhz, and channel 40 at 629 Mhz?

Answers: 9.5, 35.5, and 104.8.

5.2 PARALLEL RESONANCE

Figure 5.3 shows a perhaps more commonly used selective circuit than a series resonator. The reader should remember that because power supplies generally are full of capacitors chosen to have low impedance at 60 Hz, as far as signal is concerned, the

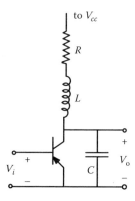

FIGURE 5.3   Parallel tuned circuit.

inductor, which also has a small resistance, runs from collector to signal ground and is in parallel with the capacitor. Then parallel resonance occurs when the capacitive admittance balances out the imaginary part of the admittance from the series combination of R and L. Thus,

$$\omega C = \text{Im}\left(\frac{1}{R+j\omega L} \times \frac{R-j\omega L}{R-j\omega L} \right) = \frac{\omega L}{R^2+(\omega L)^2} = \frac{1/\omega L}{1+(1/Q)^2}.$$

Here we see that if the Q of the circuit is fairly large, the frequency of parallel resonance is almost the same as the series resonance. What is dramatically different is the equivalent parallel resistance,

$$\frac{R^2+(\omega L)^2}{R} = R\left(1+Q^2\right),$$

which can be quite large for a large Q. It is especially convenient to have a resonator of purely parallel components because the practical resonator may also be paralleled by bias resistors. Naturally, the usual rules hold for combining parallel resistors, so the effects of shunt resistors on bandwidth are calculable. If we call the equivalent resistor R_p, the Q of a parallel resonator is

$$Q = wCR_p = R_p/wL.$$

Again, the bandwidth of a parallel resonator is given by the center frequency divided by the Q, as determined from the Q of the inductor shunted by any other resistors.

Example 5.3
An inductor of 1 μH has a Q of 40 at 455 kHz. (Figure 5.4)

   a. What capacitance is required to make a parallel resonator at 455 kHz?
   b. The capacitor is shunted by actual resistors of 50 kΩ and 470 kΩ. What will be the half-power bandwidth of the combination?

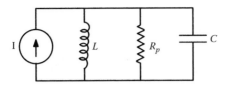

FIGURE 5.4 Purely parallel resonator circuit.

Solutions: Let's ignore the fact that the equivalent parallel inductor is increased by the factor $1 + 1/Q^2$. Hence, $C = 1/(2\pi \times 4.55 \times 10^5)^2 \times 10^{-6} = 0.122$ μF. Until we add resistors in parallel, we can still expect Q to be 40. Then,

$$R_p = \omega L Q = 4.55 \times 10^5 \times 2\pi \times 10^{-6} \times 40 = 114.4.$$

This resistance is so small that it will not be appreciably shunted by the bias resistors. Hence, the Q will still be 40 and bandwidth will be

$$B = f_0/Q = 455 \text{ kHz}/40 = 11.375 \text{ kHz}.$$

Exercise 5.4
Design an FM IF amplifier. Let the inductor be 0.5 μH and have a Q of 100 at 10.7 MHz. Find C for this resonance frequency, and find the amount of resistance that must shunt the tuned circuit for a bandwidth of 200 kHz at 10.7 Mhz.

Answers: 442 pF R_p = 3361 W for a Q of 100. Since we need Q = 53.5, the total parallel resistance must be 1798. We get this by paralleling 3361 by 3868.

Exercise 5.5
Now, a bit more challenge. In an AM receiver, we have what is called the *RF amplifier* tuned to 1600 kHz. Let us suppose we start with a 1-μH inductor that has a Q of 50 at 1.6 Mhz.

a. What capacitance is required for resonance at 1.6 Mhz?
b. By how many dBs will output be down at the image frequency 2.51 MHz?

Answers: (a) 9.9 nF, (b) 30.7 dB.

Exercise 5.6
Now, we want an FM RF amplifier at 107.9 Mhz. Design this using a 100-nH inductor with a Q of 15 and find how much rejection there is of the image at 129.3 Mhz.

Answers: 21.8 pf, 30.6 dB.

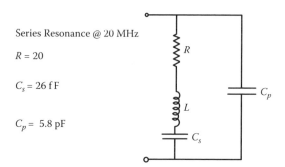

FIGURE 5.5   Equivalent circuit for piezoelectric crystal resonator.

5.3   PIEZOELECTRIC CRYSTAL RESONANCE

Piezoelectric crystals have mechanical vibrations that are highly efficient, which is to say there is almost no energy lost in a period of oscillation. The result is that there is a very high Q-series resonance. Consider Figure 5.5, showing electrical circuit parameters for one crystal.

Example 5.4

Let's calculate the equivalent value for series inductance and the resulting Q for series resonance.

Solution: $L = 1/\omega_0^2 \, C_s = 1/(4\pi \times 10^7)^2 \times 26 \times 10^{-15} = 2.436$ mH.

Hence, for series resonance, $Q = 2.436 \times 10^{-3} \times 4\pi \times 10^7/20 = 15{,}306$. This is a Q much larger than will be found anywhere else in low-frequency circuits and only can be approached in very carefully made microwave or optical resonators. For this reason, piezoelectric resonators are very useful in building signal sources in which the frequency of operation must be very accurately determined and stable.

Parallel resonance is also obtainable with piezoelectric resonators, but the selectivity, as determined by the Q, will be greatly compromised. The reason is that the frequency of resonance will be determined by the parallel capacitance C_p and the *equivalent inductance* presented just above the series resonant frequency. Since the equivalent inductance will be the series inductance *greatly reduced* by the capacitance in series with it, and series resonance is proportional to inductance whereas R is unchanged, the parallel Q is much lowered. Let's do another small calculation.

Example 5.5

It is not difficult to prove that at frequencies very near resonance, the reactive impedance of a series-resonant circuit is $2RQ\delta$, where δ is the fractional frequency shift from resonance.

Solution: We first calculate this frequency shift by requiring that the impedance of the equivalent inductor be equal to the reactance of the *parallel* capacitor, so

$2RQ\delta = 2 \times 20 \times 15{,}306 \ \delta \cong 1/4\pi \times 10^7 \times 5.8 \times 10^{-12} = 1372; \ \delta = 2.24 \times 10^{-3}.$

Thus the frequency of parallel resonance is less than 0.2% higher than the series resonance. The *equivalent* inductance is given approximately by

$$L_{eq} \cong 1/(4\pi \times 10^7)^2 \times 5.8 \times 10^{-12} = 10.92 \ \mu\text{H},$$

which is certainly much below the 2.4 mH, which teams with the series capacitance to produce a series resonance. Using the equivalent inductance but the same resistance as before, we can expect the Q to be approximately

$$Q \cong 4\pi \times 10^7 \times 10.92 \times 10^{-6}/20 = 68.6.$$

Thus, one can make parallel resonators using the piezoelectric crystal but it seems scarcely to be recommended, since the Q will be little superior to that with normal inductors and capacitors.

Exercise 5.7
Suppose one parallels the resonator above with an extra 50 pF. Estimate the shift in parallel resonant frequency and the parallel Q.

Answers: Frequency raised 0.0233%, $Q = 7.1$.

Exercise 5.8
As we saw above, one cannot "pull" the resonant frequency very far from its series value by putting rather large amounts of capacitance in parallel with the resonator. Next, the creative "what if" engineer might say, "Well, let's put in some *series* capacitance. That ought to have some effect!"

Answer: It is, "very doggone little!"

5.4 PRINCIPLES OF SINE WAVE OSCILLATORS: PHASE-SHIFT OSCILLATOR

There are times when all electrical engineers will need a source of a signal at a specific frequency. This can be supplied by what is called an *oscillator*. An oscillator may be considered an amplifier circuit that supplies an output at a certain frequency without being supplied any input signal. The trick is that one forces the amplifier to supply its own input through the application of appropriate positive feedback. One must connect a circuit called the *feedback network* to the output of the amplifier to supply exactly the input required. Suppose the amplifier boosts the input signal exactly 50 times but inverts it. Then, the appropriate feedback network reduces the output voltage by a factor of 1/50 and has a phase shift of 180°. At audio frequencies, this could be three or four cascaded networks, as shown in Figure 5.6. Suppose we write the impedance of each capacitor as jX. We can then write three mesh equations:

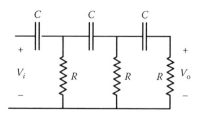

FIGURE 5.6   Feedback network for a phase-shift oscillator.

$$I_1 (R + jX) - I_2 R = V_i,$$

$$-I_1 R + I_2 (2R + jX) - I_3 R = 0, \text{ and}$$

$$-I_2 R + I_3 (2R + jX) = 0.$$

Since the output voltage depends upon I_3, we simply solve for it in terms of V_i. Then we can write the transfer function relating V_o to V_i thusly:

$$\frac{V_o}{V_i}(j\omega) = 1 - 5\left(\frac{X}{R}\right)^2 + j\left(6\frac{X}{R} - \left(\frac{X}{R}\right)^3\right).$$

The requirement of a 180° phase shift will result in the imaginary part of the transfer function required to be zero. Hence, we require $X^2 = 6\,R^2$. This results in the transfer function being equal to −29. Hence, we will have oscillations if we have an inverting amplifier with a gain of at least 29, at such a frequency that

$$\frac{1}{2\pi f C} = \sqrt{6}R.$$

Exercise 5.9
Find the value of C required for oscillations at 1 kHz, if $R = 4.7$ kΩ.

Answer: 0.0138 mF.

Exercise 5.10
Repeat Exercise 5.9 for a frequency of 4 GHz if $R = 50\ \Omega$ is used.

Answer: 0.325 pF. Besides the difficulty of getting the required passive chip components, finding an amplifier with a gain of 29 may be very difficult.

5.5   COLPITTS AND HARTLEY CONFIGURATIONS

Colpitts and Hartley configurations both use a parallel resonant circuit to provide the feedback from input to output. The difference is that the two configurations are so arranged that the voltage division is across different circuit elements. In the Hartley

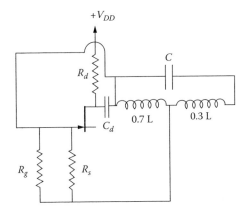

FIGURE 5.7   Hartley-connected oscillator using an field effect transistor.

configuration, the inductor is tapped so that the voltage fed back is taken across a small portion of the inductor. In the Colpitts configuration, the capacitance of the circuit is contained in a series connection of two capacitors. Consider the key elements of a Hartley circuit, shown in Figure 5.7. In this circuit, the choice of some of the elements is more critical than others; the choice of the resistors is made mainly with DC bias in mind, although one will try to keep them high enough that they do not appreciably load the resonant circuit and make its Q much lower than would simply depend upon losses in the resistor. Also, C_d is simply there to keep the left portion of the inductor from grounding the DC value of drain voltage, so it is chosen to have negligible impedance at the frequency of operation. Although it looks as though two inductors were chosen, in fact those two elements simply represent one "tapped" inductor, which is to say, a third connection is made part way from one end to the other end. The exact point of the tap is not critical; the strategy is simply to feed back a sufficient amount of the output voltage to be sure that there is more than enough to cause oscillations to build up. The choice shown implies that the gain of the FET (field effect transistor) is surely at least 2.33, and with extra gain, the waveform distorts and the loop gain requirement applies to the *fundamental* of the distorted waveform. The frequency of operation is then expected to be determined simply from the resonant frequency determined from L and C. One key consideration in the Hartley oscillator is that the frequency can be easily tweaked if there is a magnetic "slug" in the core of the form on which the inductor is wound; then a small adjustment, screwing the slug in or out of the core, will change the inductance a small amount. It might be noted that since the FET inverts the input signal as it amplifies, the required phase reversal in the feedback network is obtained by connecting the gate and drain to opposite ends of the inductor.

Figure 5.7 shows the Colpitts-connected version of an oscillator using an op-amp in the inverting configuration so as to have a small amount of inverting gain. It might be desirable to connect a potentiometer as a variable resistor for either the input resistor or the feedback resistor in the feedback path to make the gain slightly variable, so that one can make sure the gain is sufficient for oscillations to build up. Since we have shown the capacitors to be equal, theory says that oscillations should build up if

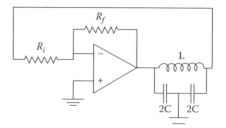

FIGURE 5.8   Op-amp-based Colpitts-connected oscillator.

feedback is such that gain is anything over unity, but Murphy's Law is perhaps more operative in building oscillators than in almost any other effort an engineer might undertake, so one must tweak the adjustable resistor until oscillations begin.

5.6   CRYSTAL-CONTROLLED OSCILLATORS

Use of the piezoelectric crystal resonance as the primary control of the frequency of operation of an oscillator can lead to a very well-defined and stable frequency. Since the Q of series resonance is very high, the circuit connections ought to capitalize on this resonance to maximize the frequency stability of the oscillations. One way in which this might be achieved is shown in Figure 5.8. Working from left to right, we first see a capacitor, which must simply be large enough that at the operation frequency, it is essentially a short circuit. Hence, we have a common-base configuration and the input of the amplifier is the emitter. The actual resistors in the circuit are there mainly for the DC biasing. The two capacitors and the inductor form a parallel resonant circuit, the main purpose of which is to filter out harmonics caused by overdriving the amplifier. It is, of course, necessary that this resonance be the same frequency as the crystal, but its Q is much lower, and the crystal is going to rule the frequency of operation.

5.7   VOLTAGE-CONTROLLED OSCILLATORS (VCO)

One does not have to go very far in the communications business before one runs across one block in a block diagram labeled VCO — the title of this section. Voltage controlled means that the frequency of oscillation depends upon the voltage fed into the VCO input. One way of doing this is fairly easy to understand. One needs to remember that a reverse-biased PN diode acts like a capacitor, the capacitance of which depends upon the amount of reverse bias. This diode could be one capacitor of a Colpitts configuration, although it might be desirable to parallel it with a fixed capacitor because using the diode alone might provide too much variation of frequency. If one is considering this method to build a so-called *direct frequency* for the FM broadcast band, one must remember that the modulation intensity required is a peak variation of frequency of ±75 kHz at a center frequency around 100 Mhz, or about 0.075%, requiring a capacitance variation of 0.15%.

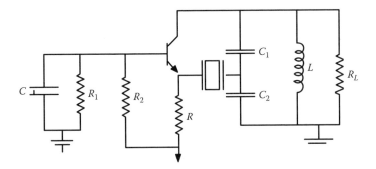

FIGURE 5.9 Crystal-controlled oscillator using feedback to noninverting input.

Example 5.6: Use of Varactor Diode in Colpitts VCO (Figure 5.10)
Data for the varactor diode from Motorola (number MMBV 2105) were inspected seeking a linear relation that would describe the variation of capacitance with reverse bias. When the negative bias is held between 1 and 2 volts, a reasonable linear approximation of the capacitance of the diode is

$$C = (28 - 5 \text{ V}) \text{ pF.}$$

Let us consider capacitance contributed for the varactor diode for each tenth of a volt from 1.0 to 2.0 volts. Combine that capacitance *in series* with 25 pF. We will then compute the resonance frequency for that capacitance in parallel with 100 nH and, from frequency changes from point to point, will get an idea of the frequency modulation sensitivity of the circuit. (See Table 5.1 for result.)

TABLE 5.1
Variation of Varactor Capacitance
Resonance and Resulting
Frequency as Bias Voltage Varies

v	C	C_{series} (pf)	f (Mhz)	Δf
1.0	23	11.979	145.416	
1.1	22.5	11.842	146.254	739
1.2	22.0	11.702	147.126	872
1.3	21.5	11.559	148.033	887
1.4	21.0	11.413	148.977	944
1.5	20.5	11.264	149.959	982
1.6	20.0	11.111	150.968	1009
1.7	19.5	10.955	152.059	1091
1.8	19.0	10.795	153.182	1123
1.9	18.5	10.632	154.352	1170
2.0	18.0	10.465	155.579	1227

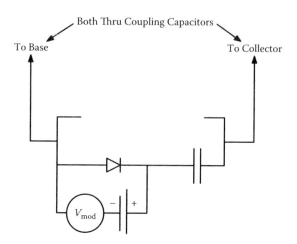

FIGURE 5.10   Varactor diode control of resonance (replaces C_1 in Figure 5.9).

Exercise 5.11 (Table 5.2)

The linearity of Δf versus voltage in the example above is not impressive. Try putting the reverse-biased diode in series with 50 pF and repeat the example above.

TABLE 5.2
Effect of a Series Capacitor
on Varactor Capacitance
tand Reseulting Resonance

v	C	C_{series} (pf)	f (Mhz)	Δf (Hz)
1.0	23	15.753	126.806	
1.1	22.5	15.517	127.766	960
1.2	22	15.278	128.762	996
1.3	21.5	15.035	129.798	1036
1.4	21	14.789	130.873	1075
1.5	20.5	14.539	131.993	1120
1.6	20	14.286	133.157	1164
1.7	19.5	14.029	134.371	1214
1.8	19	13.768	135.639	1268
1.9	18.5	13.504	136.958	1319
2.0	18	13.235	138.343	1385

This is slightly more linear than the example, but not likely to impress any but your technologically challenged grandmothers.

Example 5.7   Capacitor-Charging Determination of Frequency

Some of the more sophisticated VCOs determine their operating frequency by charging a capacitor from one voltage toward another, with each half-period ending when the capacitor voltage reaches a certain threshold. To make the calculation a *little* specific, let us choose some numbers. Suppose a capacitor voltage is being charged from −5 volts toward +5 volts. Find the frequency of cooperation in terms of the charging time constant for the capacitor if a semiperiod ends when the capacitor voltage reaches −2 volts.

Solution: From the first circuits course, we know we can write the capacitor voltage as $v_c = 5 - 10\, e^{(-t/t)}$ for all time, but in terms of period T we may write

$$-2 = 5 - 10\, e^{(-T/t)}$$

$$10\, e^{-(T/2t)} = 7$$

$$\ln(10/7) = 0.35667 = T/2t$$

$$T = 0.713\, t$$

$$f_{(operating)} = 1.44018/t.$$

Exercise 5.12
Repeat the example above for switching voltages of −3, −1, 0, 1, 2, 3 and check the variation of operating frequency with voltage.

Answers: See Table 5.3.

These results are not looking very linear in terms of frequency modulation, but

TABLE 5.3
Answers to Exercise 5.12

Switching Voltage	Operating Frequency
−3	$2.24/\tau$
−1	$0.979/\tau$
0	$0.721/\tau$
1	$0.546/\tau$
2	$0.416/\tau$
3	$0.310/\tau$

we *do* find a wide frequency variation of more than 7:1.

6 Modulation and Demodulation Circuitry

6.1 SOME FUNDAMENTALS: WHY MODULATE?

Because this chapter uses a building block approach more than do previous chapters, it may seem to be a long succession of setting up straw men and demolishing them. To some extent, this imitates the development of radio and TV, which went on for most of the twentieth century. A large number of concepts were developed as the technology advanced, and each advance made new demands upon the hardware. At first, many of these advances were made by enthusiastic amateurs who had no fear of failure and viewed radio communication the way Hillary viewed Everest — something to be surmounted "because it was there." Since about World War II, there have been increasing numbers of engineers who understood these principles and could propose problem solutions that might work the first or second time they were tried. The author fondly hopes this book will help develop a new cadre of problem solvers for the twenty-first century.

What probably first motivated the inventors of radio was the need for ships at sea to make distress calls. It may be interesting to note that the signal to be transmitted was a digital kind of thing called Morse code. Later, the medium became able to transmit equally crucial analog signals, such as a soldier warning, "Watch out!! The woods to your left are full of the abominable enemy!" Eventually, during a period without widespread military conflict, radio became an entertainment medium with music, comedy, and news, all made possible by businessmen who were convinced you could be persuaded by a live voice to buy soap, and later detergents, cars, cereals not needing cooking, and so on. The essential low- and high-frequency content of the signal to be transmitted has produced a number of problems to be solved by radio engineers.

The man on the radio, urging you to buy a pre-owned Cadillac, puts out most of his sound energy below 1000 Hz. A microphone observes pressure fluctuations corresponding to the sound and generates a corresponding voltage. Knowing that all radio broadcasting is done by feeding a voltage to an antenna, the beginning engineer might be tempted to try sending out the microphone signal directly. A big problem with directly broadcasting such a signal is that an antenna miles long would be required to transmit it efficiently. However, if the frequency of the signal is shifted a good deal higher, effective antennas become much shorter and more feasible to fabricate. This upward translation of the original message spectrum is perhaps the most crucial part of what we have come to call *modulation*. However, the necessities of retrieving the original message from the modulated signal may dictate other inclusions in the broadcast signal, such as a small or large voltage at the center, or *carrier* frequency of the modulated signal. The need for a carrier signal is dictated by what scheme is used to transmit the modulated signal, which determines important facts of how the signal can be demodulated.

More perspective on the general problem of modulation is often available by looking at the general form of a modulated signal, $f(t) = A(t)\cos\theta(t)$. If the process of modulation causes the multiplier $A(t)$ out front to vary, it is considered to be some type of *amplitude modulation*. If one is causing the angle to vary, it is said to be *angle modulation*, but there are two basic types of angle modulations. We may write

$$\theta(t) = \omega_c t + \phi.$$

If our modulation process works directly upon $\omega_c = 2\pi f_c$, we say we have performed *frequency modulation*. If instead we directly vary the phase factor $\phi(t)$, we say we have performed *phase modulation*. The two kinds of angle modulations are closely related, so that we can do one kind of operation to get the other result by proper preprocessing of the modulation signal. Specifically, if we put the modulating signal through an integrating circuit before we feed it to a phase modulator, we come out with frequency modulation. This is in fact often done. The dual of this operation is possible but is seldom done in practice. Thus, if the modulating signal is fed through a differentiating circuit before it is fed to a frequency modulator, the result will be phase modulation. However, this process offers no advantages to motivate such efforts.

6.2 HOW TO SHIFT FREQUENCY

Our technique, especially in this chapter, will be to make our proofs as simple as possible; specifically, if trigonometry proves our point, it will be used instead of the convolution theorem of circuit theory. Yet use of some of the aspects of convolution theory can be enormously enlightening to those who understand. Sometimes (as it will in this first proof) it may also indicate the kind of circuit that will accomplish the task. We will also take liberties with the form of our modulating signal. Sometimes, we can be very general, in which case it may be identified as a function $m(t)$. At other times, it may greatly simplify things if we write it very explicitly as a sinusoidal function of time $m(t) = \cos\omega_m t$. Sometimes, in the theory, this latter option is called *tone modulation*, because if one listened to the modulating signal through a loudspeaker, it could certainly be heard to have a very well-defined tone or pitch. We might justify ourselves by saying that theory certainly allows this because any particular signal we must deal with could, according the theories of Fourier, be represented as a collection, perhaps infinite, of cosine waves of various phases. We might then assess the maximum capabilities of a communication system by choosing the highest value that the modulating signal might have. In AM radio, the highest modulating frequency is typically about $f_m = 5000$ Hz. For FM radio, the highest modulation frequency might be $f_m = 19$ kHz, the frequency of the so-called FM stereo *pilot tone*. In principle, the shifting of a frequency is very simple. This is fairly obvious to those understanding convolution. One part of system theory says that multiplication of time functions leads to convolution of the spectra. Let us just multiply the modulating signal by a so-called *carrier signal*. Imagine the carrier signal "carrying" the modulating signal in the same way that homing pigeons have been used in past wars to carry a light packet containing a message from behind enemy lines to the pigeon's home in friendly territory. So, electronically, for tone modulation, we need only accomplish the product $f(t) = A \cos \omega_m t \cos \omega_c t$. We may enjoy the consequences of our assumption of tone modulation by employing trigonometric identities for the sum or difference of two angles:

$$\cos(A+B) = \cos A \cos B - \sin A \sin B \text{ and } \cos(A-B)$$

$$= \cos A \cos B + \sin A \sin B.$$

If we add these two expressions and divide by 2, we get the identity we need:

$$\cos A \cos B = 0.5\big[\cos(A+B) + \cos(A-B)\big].$$

Stated in words, we might say we got "sum and difference frequencies," but neither of the original frequencies. Let's be just a little more specific and say we started with f_m = 5000 Hz and f_c = 1 Mhz, as would happen if a radio station whose assigned carrier frequency was 1 MHz were simply transmitting a single tone at 5000 Hz. In real life this would not be done very often, but the example serves well to illustrate some definitions and principles. The consequence of the mathematical multiplication is that the new signal has two new frequencies at 995 kHz and 1005 kHz. Let's just add one modulating tone at 3,333 Hz. We would have added two frequencies at 9666.667 kHz and 1003.333 kHz. However, if this multiplication was done purely, there is no carrier frequency term present. For this reason, we say we have done a type of *suppressed carrier modulation*. Also, furthermore, we have two new frequencies for each modulating frequency. We define all of the frequencies above the carrier as the *upper sideband* and all of the frequencies below the carrier as the *lower sideband*. The whole process we have done here is called *double sideband suppressed carrier modulation*, often known by its initials, DSB-SC. Communication theory would tell us that the signal spectrum, before and after modulation with a single tone at a frequency f_m, would appear as in Figure 6.1. Please note that the theory predicts equal positive and negative frequency components. There is no deep philosophical significance to negative frequencies. They simply make the theory symmetrical and a bit more intuitive.

6.3 ANALOG MULTIPLIERS OR MIXERS

First, there is an unfortunate quirk of terminology; the circuit that multiplies signals together is usually called, in communication theory, a *mixer*. What is unfortunate is that the engineer or technician who produces sound recordings is very apt to feed the outputs of many microphones into potentiometers, the outputs of which are sent in varying amounts to the output of a piece of gear that is also called a mixer. Thus,

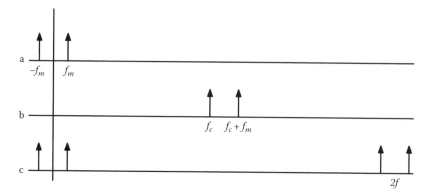

FIGURE 6.1   Unmodulated, modulated, and synchronously demodulated signal spectra. (a) Spectrum of tone-modulating signal, (b) spectrum (positive part only) of double sideband, suppressed carrier signal, (c) spectrum of synchronously-detected DSB-SC signal (except for part near $2f$).

the communication engineer's mixer multiplies and the other adds. Luckily, it will usually be obvious which device one is speaking of.

There are a number of chips available (integrated circuits) designed to serve as analog multipliers. The principle is surprisingly simple, although the chip designers have added circuitry that no doubt optimizes the operation and perhaps makes external factors less influential. The reader might remember that the transconductance f_m for a bipolar transistor is proportional to the collector current; its output is proportional to the f_m and the input voltage, so in principle one can replace an emitter resistor with the first transistor, which then controls the collector current of the second transistor. If one seeks to fabricate such a circuit out of discrete transistors, one would do well to expect a need to tweak operating conditions considerably before some approximation of analog multiplication occurs. Recommendation: Buy the chip. Best satisfaction will probably occur with a *four-quadrant multiplier*. The alternative is a *two-quadrant multiplier*, which might embarrass one by being easily driven into cutoff.

Another effective analog multiplier is alleged to be the dual-gate FET. The width of the channel in which current flows depends upon the voltage on each of two gates, which are insulated from each other. Hence, if different voltages are connected to the two gates, the current that flows is the product of the two voltages. Both devices we have discussed so far have the advantage of having some amplification, so the desired resulting signal has a healthy amplitude. A possible disadvantage may be that spurious signals one does not need may also have strong amplitudes.

Actually, the process of multiplication may be the by-product of any distorting amplifier. One can show this by expressing the output of a distorting amplifier as a Taylor series representing output in terms of input. In principle, such an output would be written

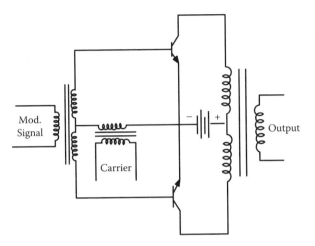

FIGURE 6.2 Balanced modulator.

$$V_0 = a_0 + a_1(v_1 + v_2) + a_2(v_1 + v_2)^2 + \text{smaller terms.}$$

One can expand $(v_1 + v_2)^2 = v_1^2 + 2v_1v_2 + 2v_2^2$, so this term yields second harmonic terms of each input plus the product of inputs one was seeking. However, the term $a_1(v_1 + v_2)$ also yielded each input, so the carrier would not be suppressed here. If it is desirable to suppress the carrier, one must resort to some sort of "balanced modulator." An *active* (meaning there is amplification provided) form of a balanced modulator is shown in Figure 6.2; failure to bias the bases of the transistors should assure that the voltage squared term is large. One will also find purely passive mixers with diodes connected in the shape of a baseball diamond with one signal fed between first and third bases, and the other from second to home plate. Such an arrangement has the great advantage of not requiring a power supply; the disadvantage is that the amplitude of the sum or difference frequency may be small.

6.4 SYNCHRONOUS DETECTION OF SUPPRESSED CARRIER SIGNALS

At this point, the reader without experience in radio may be appreciating the mathematical tricks but wondering, if one can accomplish this multiplication, can it be broadcast and the original signal retrieved by a receiver? A straightforward answer might be that multiplying the received signal by another carrier frequency signal such as $\cos \omega_c t$ will shift the signal back exactly to where it started and also up to a center frequency of twice the original carrier. This is depicted in part c of Figure 6.1. This process is called *synchronous detection*. (In the days when it was apparently felt that communications enjoyed a touch of class if one used words having Greek roots, they called it *homodyne detection*. If the reader reads a wide variety of journals, he/she may still encounter this term.) The good/bad news about synchronous detection is that the signal being used in the detector multiplication must have the exact frequency and phase of the original carrier, and such a signal is not easy to supply. One method is to send a *pilot carrier*, which is a small amount of the correct signal. The pilot tone is amplified until it is strong enough to accomplish the detection. Suppose the pilot signal reaches a high enough amplitude but is phase shifted an amount θ with respect to the original carrier. We would then, in our synchronous detector, be performing the multiplication

$$m(t)\cos\omega_c t \cos(\omega_c t + \theta).$$

To understand what we get, let us expand the second cosine using the identity for the sum of two angles,

$$\cos(\omega_c t + \theta) = \cos\omega_c t \cos\theta - \sin\omega_c t \sin\theta.$$

Hence, the output of the synchronous detector may be written

$$m(t)\cos^2\omega_c t \cos\theta - m(t)\cos\omega_c t \sin\omega_c t \sin\theta =$$

$$0.5[m(t)\cos\theta(1 - \cos 2\omega_c t) - m(t)\sin\theta\sin 2\omega_c t.$$

The latter two terms can be eliminated using a low-pass filter, and one is left with the original modulating signal, $m(t)$, attenuated proportionally to the factor $\cos\theta$, so major attenuation does not appear until the phase shift approaches 90°, when the signal would vanish completely. Even this is not totally bad news, as it opens up a new technique called *quadrature amplitude modulation* (QAM).

The principle of QAM is that entirely different modulating signals are fed to carrier signals that are 90° out of phase; we could call the carrier signals $\cos\omega_c t$ and $\sin\omega_c t$. The two modulating signals stay perfectly separated if there is no phase shift to the carrier signals fed to the synchronous detectors. The color signals in a color TV system are QAM'ed onto a 3.58-Mhz subcarrier to be combined with the black-and-white signals, after they have been demodulated using a carrier generated in synchronism with the *color burst* (several periods of a 3.58-Mhz signal), which is cleverly piggybacked onto all the other signals required for driving and synchronizing a color TV receiver.

Exercise 6.1

Prove that if the carrier signals fed to the synchronous detectors in a QAM system are both phase-shifted an amount θ, not only is the desired signal attenuated an amount $\cos\theta$, but the undesired signal will be present in an amount proportional to $\sin\theta$. The ratio of undesired to desired signal, which might be called a *crosstalk ratio*, will then be $\tan\theta$. Find the phase error for crosstalk ratios of 10, 20, 30, and 40 dB.

Answers: 17.55°, 5.71°, 1.81°, and 0.57°.

6.5 SINGLE SIDEBAND SUPPRESSED CARRIER

The alert engineering student may have heard the term *single sideband* and been led to wonder if we are proposing sending one more sideband than necessary. Of course this is true, and SSB-SC, as it is abbreviated, is the method of choice for "hams," the amateur radio enthusiasts who love to see night fall, when their low wattage signals may bounce between the earth and a layer of ionized atmospheric gasses 100 or so miles up until they have reached halfway around the world. It turns out that a little phase shift is not a really drastic flaw for voice communications, so the ham just adjusts the variable frequency oscillator being used to synchronously demodulate incoming signals until the whistles and squeals become coherent, and then he or she listens. How can one produce a single sideband? For many years it was pretty naive to say, "Well, let's just filter one sideband out!" This would have been very naive because of course one does not have textbook filters with perfectly sharp cutoffs. Recently, however, technology has apparently provided rather good *crystal lattice filters*, which are able to fairly cleanly filter the extra sideband. In general, though, the single sideband problem is simplified if the modulating signal does not go to really deep low frequencies; a microphone that does not put out much below 300 Hz might have advantages, as it would leave a transition region of 600 Hz between upper and lower sidebands in which the sideband filter could have its amplitude response "roll off" without letting through much of the sideband to be discarded. Observe Figure 6.3, showing both sidebands for a baseband signal extending only from 300 Hz to 3.0 kHz. Another method of producing a single sideband, called the *phase shift method*, is suggested if one looks at the mathematical form of just one of the sidebands resulting from tone modulation. Let us just look at a lower sideband. The mathematical form would be

$$v(t) = A\cos(\omega_c - \omega_m)t = A\cos\omega_c t\cos\omega_m t + A\sin\omega_c t\sin\omega_m t.$$

Mathematically, one needs to perform DSB-SC with the original carrier and modulating signals (the cosine terms), and with the two signals each phase-shifted 90°; the resulting two signals are then added to obtain the lower sideband. Obtaining a 90°

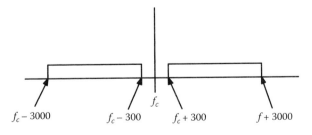

FIGURE 6.3   Double sideband spectrum for modulating signal between 300 and 3000 Hz.

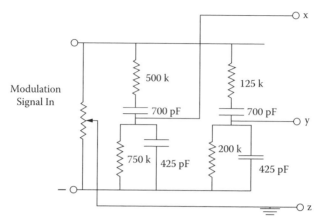

V_{xz} amd V_{yz} should be equal in magnitude but 90° apart in phase

FIGURE 6.4   Audio network for single sideband modulation.

phase shift is not difficult with the carrier, of which there is only one, but we must be prepared to handle a band of modulating signals, and it is not an elementary task to build a circuit that will produce 90° phase shifts over a range of frequencies. However, a reasonable job will be done by the circuit in Figure 6.4 when the frequency range is limited, for example, from 300 to 3000 Hz. Note that one does not modulate directly with the original modulation signal, but that the network uses each input frequency to generate two signals that are attenuated equal amounts and 90° away from each other.

Exercise 6.2
In Figure 6.4, find V_{xz} and V_y at a number of frequencies between 300 and 3000 Hz, for endpoints plus 600, 1000, 1500, 2000, and 2500 Hz.

Some answers: Numbers are fraction and phase of input voltage at each frequency.

$V_{xz} = 0.216\angle 30.58°, V_{yz} = 0.224\angle 117.65°$ at 300 Hz.

$V_{xz} = 0.216\angle -15.05°, V_{yz} = 0.224\angle 74.40°$ at 600 Hz.

$V_{xz} = 0.224\angle 59.50°, V_{yz} = 0.224\angle 148.84°$ at 3000 Hz.

6.6 AMPLITUDE MODULATION-DOUBLE-SIDEBAND-WITH-CARRIER

The budding engineer must understand that synchronous detectors are more expensive than many people can afford, and that a less expensive detection method is needed. What fills this bill much of the time is called the *envelope detector*. Let us examine some waveforms, first for DSB-SC and then for a signal having a large carrier component. Figure 6.5 shows a waveform in which similar carrier and modulating

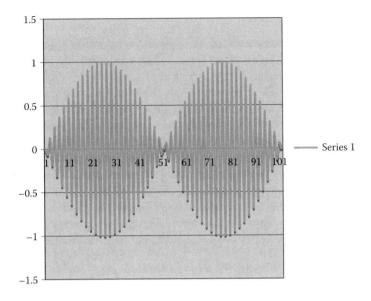

FIGURE 6.5   Double sideband suppressed carrier signal.

frequencies were chosen so that a spreadsheet plot would show the details. An ideal circuit, which we call an envelope detector, would follow the topmost excursion of the waveform sketched in Figure 6.5. The original modulating signal was a sine wave, but the topmost excursion would be a rectified sinusoid, thus containing large amounts of harmonic distortion. How can one get a waveform that will be detected without distortion by an envelope detector? What was plotted was $1.0 \cos \omega_c t \cos \omega_m t$. We suspect we must add some amount of carrier $B \cos \omega_c t$. The sum will be

$$\Phi_{AM}(t) = B \cos \omega_c t + 1.0 \cos \omega_c t \cos \omega_m t = \cos \omega_c t [B + 1.0 \cos \omega_m t].$$

This result is what is commonly called *amplitude modulation*. The following is perhaps the most useful way of writing the time function for an amplitude modulation signal having tone modulation at a frequency f_m:

$$\Phi_{AM}(t) = A \cos \omega_c t [1 + a \cos \omega_m t].$$

In this expression, we can say that A is the peak amplitude of the carrier signal that would be present if there were no modulation. The total expression inside the [ ] (brackets) can be called the *envelope* and the factor a can be called the *index of modulation*. As we have written it, if the index of modulation were >1, the envelope would attempt to go negative; this would make it necessary, for distortion-free detection, to use synchronous detection. Factor a is often expressed as a percentage, and when the index of modulation is less than 100%, it is possible to use the simplest of detectors — the envelope detector. We will look at the envelope detector in more detail a bit later.

6.6.1 MODULATION EFFICIENCY

It is good news that sending a carrier along with two sidebands makes inexpensive detection possible using an envelope detector. The accompanying bad news is that the presence of a carrier does not contribute at all to useful signal output; the presence of a carrier only leads after detection to DC, which may be filtered out at the earliest opportunity. Sometimes, as in video, the DC is needed to set the brightness level, in which case DC may need to added back in at an appropriate level.

To express the effectiveness of a communication system in establishing an output signal-to-noise ratio, it is necessary to define a *modulation efficiency*, which is simply the fraction of output power that is put into the sidebands. It is easily figured if the modulation is simply one or two purely sinusoidal tones; for real-life modulation signals, one may have to express it in quantities that are less easily visualized. For tone modulation, we can calculate modulation efficiency by simply evaluating the carrier power and the power of all sidebands. For tone modulation, we wrote

$$\Phi_{AM}(t) = A\cos\omega_c t[1 + a\cos\omega_m t],$$

which we can break into the carrier term and two sidebands as

$$A\cos\omega_c t + \frac{aA}{2}[\cos(\omega_c + \omega_m)t + \cos(\omega_c - \omega_m)t].$$

Now we have all sinusoids, the carrier, and two sidebands of equal amplitudes, so we can write the average power in terms of peak amplitudes as

$$P = 0.5\left[A^2 + 2x\left(\frac{aA}{2}\right)^2\right] = 0.5A^2\left[1 + \frac{a^2}{2}\right].$$

Then modulation efficiency is the ratio of sideband power to total power. For modulation by a single tone with modulation index a, it is

$$\eta = \frac{(aA/2)^2}{0.5A^2(1 + a^2/2)} = \frac{a^2}{2 + a^2}.$$

Of course, most practical modulation signals are not as simple as sinusoids. It may be necessary to state how close one is to overmodulating, which is to say, how close negative modulating signals come to driving the envelope negative. Besides this, what is valuable is a quantity we shall just call m, which is the ratio of average power to peak power for the modulation function. For some familiar waveforms, if the modulation is sinusoidal, $m = 1/2$. If modulation were a symmetrical square wave, $m = 1.0$; any kind of symmetrical triangle wave has $m = 1/3$. In terms of m, the modulation efficiency is

$$\eta = \frac{ma^2}{1 + ma^2}.$$

Exercise 6.3

Find the modulation efficiency for 100% modulation using a symmetrical square wave, a sine wave, and a triangle wave.

Answers: 50%, 33%, and 25%.

Exercise 6.4

If the modulated signal is kept as in Figure 6.5, how much unmodulated carrier must be added to produce modulation indices of 80% and 50%?

Answers: 1.25 sin w_ct, 2.0 sin w_ct.

Exercise 6.5

Suppose that one has an AM signal centered at 560 Mhz, with a modulation index of 0.5 that one undertakes to amplify using a Class C tuned amplifier with a $Q = 40$. What would be the modulation index at the output of the amplifier if the modulation frequency is

a. 5 kHz?
b. 3 kHz?

Answers: 40.75%, 45.98%.

6.7   ENVELOPE DETECTOR

Much detection of modulated signals — whether the signals began life as AM or FM broadcast signals or the sound or the video of TV — is done using envelope detectors. Figure 6.7 shows the basic circuit configuration. The input signal is of course as shown in Figure 6.6. It is assumed that the forward resistance of the diode is 100 ohms or less. Thus, the capacitor is small enough that it gets charged up to the peak values of the high-frequency signal, but then when input drops from the peak, the diode is reverse-biased so the capacitor can only discharge thru R. The form of this discharge voltage is given by

$$V(0)e^{(-t/RC)}.$$

The problem in AM detection is that we must have the minimum rate of decay of the voltage be at least the maximum decay of the envelope of the modulated wave. We might write the envelope as a function of time:

$$E(t) = A(1 + a\cos(\omega_m t)),$$

where A is the amplitude of the carrier before modulation and a is the index of modulation, which must be less than one for accurate results with the envelope detector.

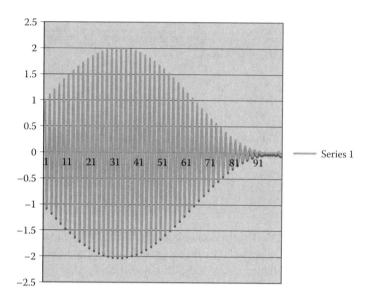

FIGURE 6.6   Fully-modulated AM signal.

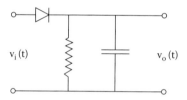

FIGURE 6.7   Simple envelope detector schematic.

Then, when we differentiate, we get

$$\frac{dE}{dt} = -\omega_m Aa \sin(\omega_m t).$$

We want this *magnitude* to be less than or equal to the maximum magnitude of the rate of decay of a discharging capacitor, which is $E(0)/RC$. For what is written as $E(0)$, we will write the instantaneous value of the envelope, and the expression becomes

$$A(1 + \cos(\omega_m t) \geq RC(\omega_m aA \sin(\omega_m t)).$$

The *A*s cancel, and we have

$$RC \leq \frac{1 + a\cos(\omega_m t)}{\omega_m a \sin(\omega_m t)}.$$

Our major difficulty occurs when the right-hand side has its minimum value. If we differentiate with respect to $\omega_m t$, we get

$$\frac{\omega_m a \sin(\omega_m t) \times (-\omega_m a \sin(\omega_m t)) - (1 + a\cos(\omega_m t))\omega_m^2 a \cos(\omega_m t)}{(\omega_m a \sin(\omega_m t))^2}.$$

We set the numerator equal to zero to find its maximum. We then have

$$-(\omega_m a)^2 \left[\sin^2(\omega_m t) + \cos^2(\omega_m t) \right] - a(\omega_m)^2 \cos(\omega_m t) = -(\omega_m a)^2 - a\omega_m^2 \cos(\omega_m t).$$

Hence, the maximum occurs when $\cos(\omega_m t) = -a$, and, of course, by identity, at that time,

$$\sin(\omega_m t) = \sqrt{1 - a^2}.$$

Inserting these results into our inequality for the time constant RC, we have

$$RC \le \frac{1 - a^2}{\omega_m a \sqrt{1 - a^2}} = \frac{\sqrt{1 - a^2}}{\omega_m a}.$$

Example 6.1
Suppose we say 2000 Hz is the main problem in our modulation scheme, our modulation index is 0.5, and we chose $R = 10k$ to make it large compared to the diode forward resistance, but not too large. What should be the capacitor C?

Solution: We use the equality now and get $C = 13.8$ nF.

Exercise 6.6
In demodulating a video picture, a maximum modulation of 8% occurs at 15,750 Hz. Our resistor is 20 k. Find the maximum size for the capacitor.

Answer: 6.29 nF.

6.8 ENVELOPE DETECTION OF SSB USING INJECTED CARRIER

Single sideband (SSB) is a very forgiving medium, it might be said. Suppose that one were attempting synchronous detection using a carrier that was off by a Hertz or so, compared to the original carrier. Since synchronous detection works by producing sum and difference frequencies, a 1.0-Hz error in carrier frequency would produce a 1.0-Hz error in the detected frequency. Since SSB is mainly used for speech, it would be challenging indeed to find anything wrong with the reception of a voice one has only ever heard over a static-ridden channel. Similar results would also be felt in the following, where we add a carrier to the sideband and find that we have AM, albeit with a small amount of harmonic distortion.

Example 6.2

Starting with just an upper sideband $B\cos(\omega_c + \omega_m)t$, let us add a carrier term $A\cos\omega_c t$, manipulate the total, and prove that we have an envelope to detect. First we expand the sideband term as

$$\phi_{SSB}(t) = B(\cos\omega_c t \cos\omega_m t - \sin\omega_c t \sin\omega_m t).$$

Adding the carrier term $A\cos\omega_c t$ and combining like terms, we have

$$\phi(t) = (A + B\cos\omega_m t)\cos\omega_c t - B\sin\omega_c t \sin\omega_m t.$$

In the first circuits class, we see that if we want to write a function of one frequency in the form $E(t)\cos(\omega_c t + \text{phaseangle})$, the amplitude of the multiplier E is the square root of the squares of the coefficients of $\cos(\omega_c t)$ and $\sin(\omega_c t)$. Thus

$$E(t) = \sqrt{(A + B\cos\omega_m t)^2 + (B\sin\omega_m t)^2}$$

$$= \sqrt{A^2 + 2AB\cos\omega_m t + B^2(\cos^2(\omega_m t) + \sin^2(\omega_m t))}.$$

Now, of course the coefficient of B^2 is unity for all values of $\omega_m t$. We find that the best performance occurs if $B \ll A$. Then we would have our expression for the envelope (which is hence detectable using an envelope detector):

$$E(t) = \sqrt{A^2 + B^2 + 2AB\cos\omega_m t} = \sqrt{A^2 + B^2}\sqrt{1 + \frac{2AB}{A^2 + B^2}\cos\omega_m t}.$$

Our condition that $B \ll A$ allows us to say that the coefficient of $\cos\omega_c t$ is really small compared to unity. We use the binomial theorem to approximate the second square root:

$$(1 + x)^{1/2} \approx 1 + x/2 - x^2/8$$

when $x \ll 1$. Using our approximation,

$$x = \frac{2B}{A}\cos\omega_m t.$$

In our expansion, the x term is the modulation term we were seeking; the x^2 term contributes second harmonic distortion. Using the various approximations, and stopping after we find the second harmonic (other harmonics will be present, of course, but in decreasing amplitudes), we have:

$$\text{Detected } f(t) = \frac{B}{A}\cos\omega_m t - \frac{B^2}{2A}\cos^2\omega_m t.$$

When we use trig identities to get the second harmonic, we get another factor of one-half; the ratio of the detected second harmonic to the fundamental is thus $(1/4)(B/A)$. Thus, for example, if B is just 10% of A, the second harmonic is only 2.5% of the fundamental.

Exercise 6.7
Use the approximate formula derived above to find the conditions for a 5% second harmonic and a 1% second harmonic.

Answers: $B = 0.2\,A$, $B = 0.05\,A$.

6.9 DIRECT VERSUS INDIRECT MEANS OF GENERATING FM

Let us first remind ourselves of the basics regarding FM. We can write the time function in its simplest form as

$$\phi_{FM}(t) = A\cos(\omega_c t + \beta\sin\omega_m t).$$

Now, the alert reader might be saying, "Hold on! That looks a lot like phase modulation!" If $\beta = 0$, the phase would increase linearly in time, as an unmodulated signal, but the modulated signal gets advanced or retarded a maximum of β. One needs to remember the definition of instantaneous frequency, which is

$$f_i = \frac{1}{2\pi}\frac{d}{dt}(\omega_c t + \beta\sin\omega_m t) = \frac{1}{2\pi}(2\pi f_c + \beta 2\pi f_m\cos\omega_m t) = f_c + \beta f_m\cos\omega_m t.$$

Thus we can say that instantaneous frequency departs from the carrier frequency by a maximum amount βf_m, which is the so-called *frequency deviation*. This has been specified as a maximum of 75 kHz for commercial FM radio but 25 kHz for the audio of TV signals.

Now, certainly, the concept of directly generating FM has an intellectual appeal to it. The problems of direct FM are mainly practical; if the very means of putting information onto a high-frequency carrier is in varying the frequency, it perhaps stands to reason that the center value of the operating frequency will not be well defined. Direct FM could be accomplished as in Figure 6.8. However, Murphy's Law would be very dominant and one might expect the center frequency to drift continually in one direction all morning and the other way all afternoon, or the like. This system is sometimes stabilized by having an FM detector called a *discriminator* tuned to the desired center frequency, so that the output would be positive if the frequency got high and negative for low frequency. Thus, instantaneous output could be used as an error voltage with a long time constant to push the intended center frequency toward the center, whether the average value was above or below.

The best-known method of indirect FM gives credit to the man who, more than any other, saw the possibilities of FM and that its apparent defects could be exploited for superior performance — Edwin Armstrong. He started with a crystal-stabilized oscillator around 100 kHz, from which he obtained a 90° phase-shifted version. A

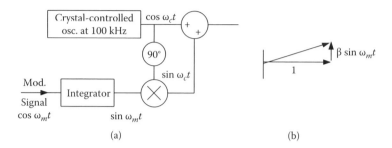

FIGURE 6.8   (a) Crystal-stablized phase modulator, (b) phasor diagram.

block diagram of this early part of the Armstrong modulator is shown in Figure 6.8. The modulating signal is passed through an integrator before it goes into an analog multiplier, to which is also fed the phase-shifted version of the crystal-stabilized signal. Thus, we feed $\cos(\omega_c t)$ and $\sin(\omega_c t)\sin\omega_m t$ into a summing amplifier. The phasor diagram shows the two signals with $\cos(\omega_c t)$ as the reference. There is a small phase shift given by $\tan^{-1}(\beta\sin\omega_m t)$, where β gives the maximum amount of phase shift as a function of time. To see how good a job we have done, we need to expand $\tan^{-1}(x)$ in a Taylor series. We find that

$$\tan^{-1}(x) \approx x - (x)^3/3 + (x)^5.$$

We see that we have a term proportional to the modulating signal (x) and others that must represent odd-order harmonic distortion, if one accounts for the fact that we have resorted to a subterfuge, using a phase modulator to produce frequency modulation. Assuming that finally our signal goes through a frequency detector, we find that the amount of the third harmonic as a fraction of the signal output is $\beta^2/4$. Now, in frequency modulation, the maximum amount of modulation permitted is expressed in terms of frequency deviation, an amount of 75 kHz. The relation between frequency deviation and maximum phase shift is $\Delta f = \beta f_m$, where Δf is the frequency deviation, β is the maximum phase shift, and f_m is the modulation frequency. Since maximum modulation is defined in terms of β, the maximum value of β permitted will correspond to the minimum modulation frequency. Let us do some numbers to illustrate this problem.

Example 6.3
Suppose we have a high-fidelity broadcaster wishing to transmit bass down to 50 Hz with a maximum third harmonic distortion of 1%. Find the maximum values of β and frequency deviation Δf.

Solution: We have $\dfrac{\beta^2}{4} = 0.01$. Solving for β, we get $\beta = 0.2$. Then $\Delta f = 0.2 \times 50$ Hz = 10 Hz.

One can recall that the maximum value of frequency deviation allowed in the FM broadcast band is 75 kHz. Thus, use of the indirect modulator has given us much lower frequency deviation than is allowed, and clearly some kind of desperate measures are required. Such are available, but they do complicate the process greatly. Suppose we feed the modulated signal into an amplifier that is not biased for low distortion, that is, its Taylor series looks like

$$f(x) = a_1 x + a_2 x^2 + a_3 x^3.$$

Now, the squared term leads to the second harmonic, the cubed one gives the third harmonic, and so on. The phase-modulated signal looks like $A \cos(\omega_c t + \beta \sin \omega_m t)$, and the term $a_2 x^2$ not only doubles the carrier frequency but also the maximum phase shift β. Thus, starting with the rather low carrier frequency of 100 kHz, we have a fair amount of multiplying room before we arrive in the FM broadcast band of 88 to 108 MHz. Unfortunately, we may need different amounts of multiplication for the carrier frequency than we need for the depth of modulation. Let's carry on our example and see the problems that arise. First, if we wish to go from $\Delta f = 10$ Hz to 75,000, that leads to a total multiplication of 75,000/10 = 7500. The author likes to say we are limited to frequency doublers and triplers. Let's use as many triplers as possible; we divide the 7500 by 3 until we get close to an exact power of 2:

$$7500/3 = 2500, \ 2500/3 \cong 833, \ 833/3 = 278, \ 278/3 \ ^a \approx 93, \ 93/3 = 31,$$

which is very close to $32 = 2^5$. So, to get our maximum modulation index, we need five triplers and five doublers. However, 7500×0.1 Mhz = 750 Mhz, and we have missed the broadcast band by about 7 times. One more thing we need is a mixer, after a certain amount of multiplication. Let's use all the doublers and one tripler to get a multiplication of $32 \times 3 = 96$, so the carrier arrives at 9.6 Mhz. Suppose our final carrier frequency is 90.9 Mhz, and since we have remaining to be used a multiplication of $3^4 = 81$, what comes out of the mixer must be 90.9 MHz/81 = 1.122 MHz. To obtain an output of 1.122 Mhz from the mixer, with 9.6 MHz going in, we need a local oscillator of either 10.722 or 8.478 MHz. Note that this local oscillator also needs a crystal control, or the eventual carrier frequency will wander about more than is allowed.

Exercise 6.8
The station engineer decides that a maximum third harmonic distortion of 4% at 50 Hz is permissible. What would be the key numbers for an Armstrong modulator?

Answer: Total multiplication of 3750 needed, or five triplers and four doublers. Perhaps, the first bit of multiplication might be $3^4 = 81$. That leaves a multiplication of 48, so one needs 1.89375 Mhz out of the mixer; thus the crystal-controlled local oscillator must be at 9.99375 MHz or 6.20625 Mhz.

Exercise 6.9
Meanwhile, the cynical engineer designing an Armstrong modulator for the station at 107.9 Mhz says, "Oh, what the hey. Our listeners are kids without decent woofers any way. Let's design for 5% third harmonic at 100 Hz!" What are his or her numbers?

Answers: $\beta = 0.447$, so $\Delta f = 44.7$ Hz, total multiplication $=1677$, approximated by $3^3 \times 2^8=$. Before mixer, try $27 \times 4 = 108$, so 10.8 Mhz could be one input into the mixer. Output must be 6.47375 MHz, so LO must be 17.54 Mhz, or at 4.56 Mhz.

6.10   QUICK-AND-DIRTY FM SLOPE DETECTION

A method of FM detection that is barely respectable, but surprisingly effective, is called *slope detection*. The principle is to feed an FM signal into a tuned circuit, not right at the resonant frequency, but rather somewhat off the peak. Therefore, the frequency variations due to the modulation will drive the signal up and down the resonant curve, producing simultaneous amplitude variations that can then be detected using an envelope detector. Let us just take a case of FM and a specific tuned circuit and find the degree of AM.

Example 6.3
We have an FM signal centered at 10.7 Mhz, with frequency deviation of 75 kHz. We have a purely parallel resonant circuit with a $Q = 30$, and a resonant frequency such that 10.7 Mhz is at the lower half-power frequency. Find the output voltage for $\Delta f = +75$ kHz and for -75 kHz.

Solution: When we operate close to resonance, adequate accuracy is given by

$$V_o = \frac{V_i}{1 + j2Q\delta},$$

where δ is the fractional shift of frequency from resonance. If now 10.7 Mhz is the lower half-power point, we can say that $2Q\delta = 1$. Hence $\delta = 1/(2 \times 30) = (f_0 - 10.7 \text{ Mhz})/f_0$; $f_0 = 10.881$ Mhz.

Now we evaluate the transfer function at 10.7 Mhz $\pm$ 75.kHz. We defined it as 0.7071 at 10.7 Mhz. For 10.7 + 0.075 Mhz, $\delta = (10.881 - 10.775)/10.881 = 9.774 \times 10^{-3}$, and the magnitude of the transfer function is $|1/(1 + j60\delta)| = 0.8626$. Since the value was 0.7071 for the unmodulated wave, the modulation index in the positive direction would be $(0.8624 - 0.7071)/ 0.7071 = 0.2196$ or 21.96%. For $(10.7 - 0.075)$ Mhz, $\delta = (10.881 - 10.625)/10.881 = 0.02356$, and the magnitude of the transfer function is $|1/(1 + j60)| = 0.5775$. The modulation index in the negative direction is $(0.7071 - 0.5775)/0.7071 = 18.32\%$. So, the modulation index is not the same for positive as for negative indices. The consequence of such asymmetry is that this process will be subject to harmonic distortion, which is why it is not quite respectable.

Exercise 6.10
Repeat the example above if the Q of the circuit is 50.

Answers: The amplitude modulation index is 35.22% up and 28.1% down, so it looks as though we have even worse curvature and harmonic distortion.

Exercise 6.11

Repeat the example above except that one places 10.7 Mhz at the −6.02 dB point.

Answers: The amplitude modulation index is 20.56% up and 15.34% down.

6.11 LOWER DISTORTION FM DETECTION

We will assume that the reader has been left wanting an FM detector that has much better performance than the slope detector. A number of more complex circuits have a much lower distortion level than the slope detector. One, called the *balanced FM discriminator*, is shown in Figure 6.9. Basically, we may consider that the circuit contains two *stagger-tuned* resonant circuits, that is, they are tuned equidistant on opposite sides of the center frequency, connected back to back. The result is that the nonlinearity of the resonant circuits balances each other out, and the FM detection can be very linear. The engineer designing an FM receiving system can easily access such performance; all he or she must do is spend the money to obtain high-quality components.

6.11.1 PHASE-LOCKED LOOP

The phase-locked loop is an assembly of circuits or systems that can perform a number of functions to accomplish several useful operations, any one or more of which might be capitalized upon. If one looks at a simple block diagram, one will see something like Figure 6.10. Thus, one function that will always be found is called a *voltage-controlled oscillator*; this term means that there is an oscillator that would run freely at some frequency, but if a nonzero DC voltage is fed into a certain input, the frequency of oscillation will shift to one determined by that input voltage. Another function one will always find (although the nomenclature might vary somewhat) is *phase comparison*. The phase *comparator* will usually be followed some kind of low-pass filter. Of course, if a comparator is to fulfill its function, it requires two inputs, the phases of which it compares. This operation might be accomplished in various ways; however, one method that might be understood from previous discussions is the *analog multiplier.* Suppose an analog multiplier receives the inputs cos(ωt) and sin($\omega t + \phi$); their product has a sine and a cosine. Now, a trigonometric identity involving these terms is

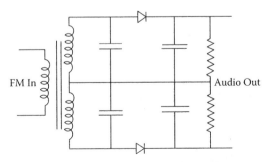

FIGURE 6.9 Balanced FM discriminator.

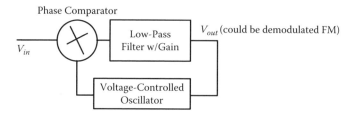

FIGURE 6.10 Basic phase-locked loop.

$$\sin A \cos B = 0.5 \big[\sin(A+B) + \sin(A-B) \big].$$

Thus the output of a perfect analog multiplier will be 0.5 [sin(2ωt + ϕ) + sin(ωt)]. A low-pass filter following the phase comparator is easily arranged; therefore, one is left with a DC term, which, if it is fed to the VCO in such a polarity as to provide negative feedback, will "lock" the VCO to the frequency of the input signal with a fixed phase shift of 90°.

Phase-locked loops (abbreviated PLL) are used in a wide variety of applications. Many of the applications are demodulators of one sort or another, such as synchronous detectors for AM, basic FM, FM-stereo detectors, and in very precise oscillators known as *frequency synthesizers*. One of the early uses seemed to be the detection of weak FM signals, where it can be shown that they extend the threshold of useable weak signals a bit. This latter facet of their usefulness seems not to have made a large impact, but the other aspects of PLL usefulness are very commonly seen.

6.12 DIGITAL MEANS OF MODULATION

The sections immediately preceding have been concerned with rather traditional analog methods of modulating a carrier. While the beginning engineer can expect to do little or no designing in analog communication systems, they serve as an introduction to the digital methods that most certainly will dominate the design work early in the twenty-first century. Certainly analog signals will continue to be generated, such as speech, music, and video; however, engineers are finding it so convenient to do digital signal processing that many analog signals are digitized, processed in various performance-enhancing ways, and only restored to analog format shortly before they are fed to a speaker or picture tube. Digital signals can be transmitted in such a way as to use extremely noisy channels. Using a surprisingly durable exploration vehicle, in late 2006 the nightly news brought us video of the Martian landscape. The analog engineer would be appalled to know the number representing the traditional signal-to-noise ratio for the Martian signal. The detection problem is greatly simplified because the digital receiver does not need at each instant to try to represent which of an infinite number of possible analog levels is correct; it simply asks, was the signal sent a one or a zero? That is simplicity.

Several methods of digital modulation might be considered extreme examples of some kind of analog modulation. Recall amplitude modulation. The digital rendering of AM is called *amplitude shift keying* (ASK). What this might look like on

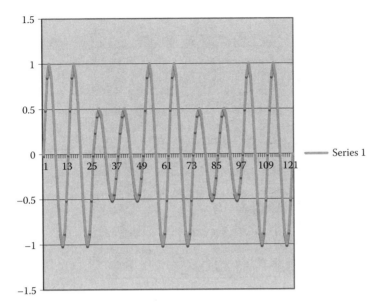

FIGURE 6.11 Amplitude shift keying.

an oscilloscope screen is shown in Figure 6.11. For example, we might say that the larger amplitude signals represent the logic ones and smaller amplitudes represent logic zeroes. Thus, we have illustrated the modulation of the data stream 10101. If the intensity of modulation were carried to the 100% level, the signal would disappear completely during the intervals corresponding to zeroes. The 100% modulation case is sometimes called *on-off keying* (OOK). The latter case has one advantage if this signal were nearly obscured by large amounts of noise: It is easiest for the digital receiver to distinguish between ones and zeroes if the difference between them is maximized. That is, however, only one aspect of the detection problem. It is also often necessary to know the timing of the bits, and for this one may use the signal to synchronize the oscillator in a phase-lock loop. If for 50% of the time there is zero signal by which to be synchronized, the oscillator may drift significantly. In general, a format for digital modulation in which the signal may vanish utterly at intervals is to be adopted with caution and with full cognizance of the ones sync problem. Actually, amplitude shift keying is not considered a very high-performance means of digital signaling, in much the same way that AM is not greatly valued as a quality means of analog communication. Frequency shift keying and phase shift keying are the methods most often used.

6.12.1 FREQUENCY SHIFT KEYING

Frequency shift keying (FSK) may be used in systems having very little to do with high data rate communications; for years it has been the method used in the simple modems one first used to communicate with remote computers. For binary systems, one just sent a pulse of one frequency for a logic one and a second frequency for a

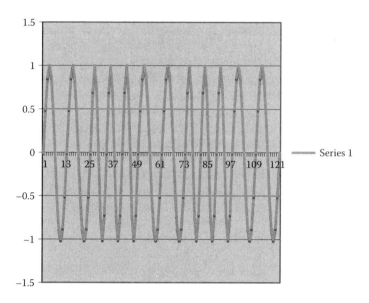

FIGURE 6.12   Frequency shift keying.

logic zero. If one is communicating in a noisy environment, the two signals should be orthogonal, which means that the two frequencies used should be separated by at least the data rate. At first the modem signals were sent over telephone lines, which were optimized for voice communications and were rather limited for data communication. Suppose we consider that for ones we send a 1250-Hz pulse and for zeroes we send a 2250-Hz pulse. In a noisy environment one ought not to attempt sending more than 1000 bits per second (note that 1000 Hz is the exact difference between the two frequencies being used for FSK signaling). Let us instead send at least 250 bps; 12 milliseconds of a 101 bit stream is seen in Figure 6.12. It is not too difficult to imagine a way to obtain FSK. Assuming one does have access to a VCO, one simply feeds it two different voltage levels for ones and for zeroes. The VCO output is the required output.

6.12.2 PHASE SHIFT KEYING

Probably the most commonly used type of digital modulation is some form of phase shift keying. One might simply say there is a carrier frequency f_c, and that logic zeroes will be represented by $-\sin(2\pi f_c t)$ and logic ones by $+\sin(2\pi f_c t)$. If the bit rate is 40% of the carrier frequency, the data stream 1010101010 might look like Figure 6.13. In principle, producing binary phase shift keying ought to be fairly straightforward if one has the polar NRZ (non-return to zero, meaning a logic one could be a constant positive voltage for the duration of the bit, with zero being an equal negative voltage) bit stream. If the bit stream and a carrier signal are then fed into an analog multiplier, the output of the multiplier could indeed be considered $\pm\cos\omega_c t$, and the modulation is achieved.

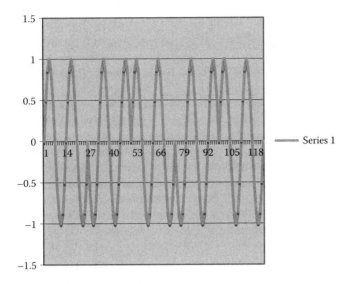

FIGURE 6.13   Phase shift keying.

6.13   CORRELATION DETECTION

Many years ago, the communications theorists came up with the idea that one could build a *matched filter*, that is, a special filter designed with the bit waveform in mind that would startlingly increase the signal-to-noise ratio of the detected signal. Before long, a practically minded communications person had the bright idea that a correlation detector would do the job, at least for rectangular bits. For some reason, as one explains this circuit, one postulates two signals, $s_1(t)$ and $s_2(t)$, which represent, respectively, the signals sent for logic ones and zeroes. The basics of the correlation detector are shown in Figure 6.14. A key consideration in the operation of the correlation detector is bit synchronization. It is crucial that the signal $s_1(t)$ be lined up perfectly with the bits being received. Then, the top multiplier "sees" $\sin(\omega_c t)$ coming in one input and $\pm\sin(\omega_c t)$ + noise coming in the other, depending upon whether a one

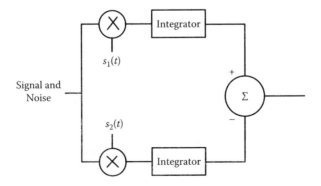

FIGURE 6.14   Correlation detector.

or a zero is being received. If it happens that a one is being received, the multiplier is asked to multiply $\sin(\omega_c t)(\sin(\omega_c t + \text{noise}))$:

$$\text{Of course, } \sin^2 \omega_c t = \tfrac{1}{2}(1 + \cos(2\omega_c t)).$$

In the integrator, this is integrated over one bit duration, giving a quantity said to be the energy of one bit. The integrator might also be considered to have been asked to integrate $n(t)\sin\omega_c t$, where $n(t)$ is the noise signal. However, the nature of noise is that there is no net area under the curve of its waveform, so considering integration to be a summation, the noise output out of the integrator would simply be the last instantaneous value of the noise voltage at the end of a bit duration, whereas the signal output was bit energy if the bit synchronization is guaranteed. Meanwhile, the output of the bottom multiplier was the negative of the bit energy, so with the signs shown, the output of the summing amplifier is twice the bit energy. Similar reasoning leads to the conclusion that if the instantaneous signal being received were a zero, the summed output would be minus twice the bit energy. It takes a rather substantial bit of theory to show that the noise output from the summer is noise spectral density. The result may be summarized that the correlation detector can "pull a very noisy signal out of the mud." And we should assert at this point that the correlation detector can perform wonders for any one of the methods of digital modulation mentioned up to this point.

6.14 DIGITAL QAM

Once the engineer has produced carrier signals that are 90° out of phase with each other, there is no intrinsic specification that the modulation must be analog, as is done for color TV. As a start toward extending the capabilities of PSK, one might consider that one sends bursts of several periods of $\pm\cos(\omega_c t)$ or $\pm\sin(\omega_c t)$. This is sometimes called *4-ary transmission*, meaning that there are 4 different possibilities of what might be sent. Thus, whichever of the possibilities is sent, it may be considered to contain 2 bits of information; it is a method by which more information may be sent without demanding any more bandwidth, since the duration of the symbol being sent may be no longer or shorter than it was when one was doing binary signal sending, for example, simply $\pm\cos(\omega_c t)$. This idea is sometimes represented in a *constellation*, which, for the case we just introduced, would look like part (a) of Figure 6.15. However, what is done more often is shown in Figure 6.15(b), where it could be said that one is sending

$$\pm\cos(\omega_c t + 45°) \text{ or } \pm\cos(\omega_c t + 135°).$$

It seems as though this may be easier to implement than the case in part (a); however, the latter scheme lends itself well to sending 4 bits in a single symbol, as in Figure 6.15(c). Strictly speaking, one might consider Figure 6.15(a) to be the constellation for a 4-ary PSK. This leads to the implication that one could draw a circle with "stars" spaced 45° apart on it and one would have the constellation for an 8-ary PSK.

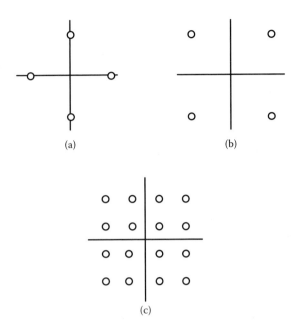

FIGURE 6.15 (a) 4-ary, elementary; (b) 4-ary, more practical; (c) 16-ary constellation.

The perceptive or well-informed reader might have the strong suspicion that crowding more points on the circle makes it possible to have more errors in distinguishing one symbol from adjacent ones, and he or she would be correct in this suspicion. Hence, M'ary communication probably more commonly uses Figure 6.15(b) or (c), which should be considered forms of QAM.

We will consider some of the performance aspects of digital modulation in the following chapter, when we consider their noise rejection properties.

7 Thermal Noise and Amplifier Noise

7.1 THERMAL NOISE

Long before the modern-day engineer was born, Nyquist thought of the electrons in a conductor as behaving like a gas and applied his understanding of thermodynamics to predicting the nature of random voltages generated from the electrons jiggling around. Subsequently, van der Ziel added the first term, which is a linear function of frequency. The result is expressed most usefully in terms of a power spectral density, which might be observed by instruments having infinite input impedance but finite bandwidth, connected across resistance R. The result is often written as

$$P(f) = 2R\left(\frac{h|f|}{2} + \frac{h|f|}{e^{|f|/kT} - 1}\right),$$

where $|f|$ assumes that one will be integrating over positive and negative frequencies with equal contributions and the "2" out front assumes such integration.

Since the integrand *is* an even function, it is best to just multiply by another 2 and integrate over the positive frequency. The letter "h" is Planck's constant,

$$6.625 \times 10^{-34} \text{ J-sec,}$$

with the result that the first term in the bracket is negligible until one gets to a very high frequency, such as visible or ultraviolet light. The letter "k" is Boltzmann's gas constant,

$$1.38 \times 10^{-23} \text{ K,}$$

where K stands for Kelvin degrees and T is Kelvin temperature, which the reader may recall is Celsius temperature + 273. Now, at low frequencies, further approximations are possible (and very welcome). It may be recalled that one can expand the exponential function in an infinite series, but that one can keep just two terms if the exponent is small compared with one. Suppose we decide that "small compared to one" means one-tenth or less. Let's assume temperatures near a typical (wintertime) room temperature of 290 K, and find what maximum frequency this implies:

$$\frac{h|f|}{kT} = 0.1; \quad |f| = \frac{0.1 \times 1.38 \times 10^{-23} \times 290}{6.2 \times 10^{-34}} = 645 \text{ GHz.}$$

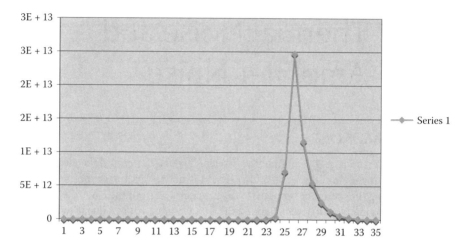

FIGURE 7.1   Noise spectrum—noise power versus frequency in half-powers of ten.

One will never approach such frequencies except in the optics lab, so for most frequencies with which we would want to do noise calculations, we *may* write e^x as $1 + x$, to a good approximation. Hence, we change the spectral density expression to

$$P(f) = \frac{4Rh|f|}{1 + \dfrac{h|f|}{kT} - 1} = \frac{4Rh|f|}{\dfrac{h|f|}{kT}} = 4\ \mathrm{kTR}.$$

Thus, we see that at low frequencies, the actual frequency is removed from the power spectral density expression, and it is a constant independent of frequency. In the early days when physicists were learning about light, they noted from passing sunlight through a prism that apparently white light is actually made up of a spectrum of colors. Similarly, at low frequencies, thermal noise has the same amount of noise power in each Hertz of bandwidth, so thermal noise is said to be "white noise." In Figure 7.1 we have plotted the thermal noise power density versus powers of ten in frequency to dramatize this behavior, at suboptical frequencies.

The result is often stated that if one connected a voltmeter of infinite input impedance, but a sharp cutoff at a bandwidth B, one would observe

$$v^2 = 4\ \mathrm{kTRB}.$$

Example 7.1
Find the thermal noise voltage that one should observe across a 250-k resistor at room temperature, using a voltmeter with a bandwidth of 1 Mhz.

Solution: $v = \sqrt{4 \times 1.38 \times 10^{-23} \times 290 \times 2.5 \times 10^5 \times 10^6} = 63.3\,\mu v.$

Before the student decides to verify this number experimentally, it should be hastily mentioned that in the typical lab or industrial setting, many other factors besides thermal effects are generating interfering voltages, and that if one is determined to try the experiment, one should be sure to try it in a *screened room*, in which most interfering voltages, except thermally generated ones, are screened out.

7.1.1 AVAILABLE NOISE POWER

In circuit theory, the word "available" means the amount of power that will be delivered to a load that is impedance-matched to the source. This means, if we could connect a noiseless resistor equal to R across the noisy R, what power would be observed? First, one should realize that one has constructed a 2:1 voltage divider, so the voltage across the lossless resistor would be obtained from

$$v^2 = \left(\tfrac{1}{2}\right)^2 \times 4 \ kTRB = kTRB.$$

If one wants to know the available noise *power*, one divides by R, getting

$$N_{avail} = kTB.$$

Example 7.2
One can generally expect that laboratory signal generators have their output impedances at room temperature. What available noise power into a matched load can one expect from a generator with 1.0 MHz bandwidth?

Solution: $kTB = 1.38 \times 10^{-23} \times 290 \times 10^6 = 4 \times 10^{-15}$ watts.

Example 7.3
The main departures from room temperature of sources of noise come from antennas. In the AM band, source temperatures of 10^7 K are not unusual. In the FM band, one should expect source temperatures of about 1000 K. However, UHF TV may experience source temperatures around 100 and around 2 to 4 GHz, and directive antennas pointed straight up in the air may see temperatures of 10 or 20 K. This has very happy implications for satellite TV. Let us find the available noise power in the AM band where the bandwidth is 10^4 Hz.

Solution: $kTB = 1.38 \times 10^{-23} \times 10^7 \times 10^4 = 1.38 \times 10^{-12}$ watt, or 1.38 pw.

Exercise 7.1
Find the available noise power into an FM receiver with a bandwidth of 200 kHz, matched to an antenna that sees a sky temperature of 2000 K.

Answer: 5.52 femtowatts = 5.52×10^{-15} watts.

Exercise 7.2
What is the available noise power into a satellite TV receiver with a bandwidth of 12 MHz if the antenna is pointed in a direction where the temperature is 50 K?

Answer: 8.28 fw.

7.2 AMPLIFIER NOISE FIGURE AND EXCESS TEMPERATURE

In Candide's "best of all possible worlds," the noise out of an amplifier would be simply the power gain of the amplifier times the noise being fed into the amplifier. In this real world, the output noise is always a little higher. This leads to the definition of a quality factor for an amplifier, which we will call its *noise factor*, which is defined at a standard room temperature of 290 K as the ratio of the amplifier's input signal-to-noise ratio divided by its output signal-to-noise ratio. This is given the symbol F and may be written

$$F = \frac{(S/N)_{in}}{(S/N)_{out}} = \frac{S_{in}/kT_oB}{GS_{in}/(GkT_oB + N_{excess})} = \frac{(GkT_oB + N_{excess})}{GkT_oB}.$$

It should be noted that the noise factor (and its related quality factor, the *noise figure*) is only useful if the source driving the amplifier is at so-called *room temperature* (T_o = 290 K). Otherwise, an alternate concept, the *excess temperature* of the amplifier, is very useful. It is implied that excess noise power results from higher than expected temperature at the input, and one can write

$$N_{excess} = GkT_{excess}B,$$

hence one can write

$$F = 1 + T_e/T_o.$$

Indeed, there will be times when one thinks the only use of the noise factor is that it enables one to find the excess temperature. Note also that the noise factor is always >1.

Example 7.4
Find the noise factor corresponding to an excess temperature of 58 K.

Solution: $F = 1 + \dfrac{58}{290} = 1.2$.

Originally, the IEEE thought that if one took 10 $\log_{10}(F)$, it would be known as "noise figure." In some texts, it receives the symbol F_{dB}.

Example 7.5
Find the noise figure corresponding to an excess temperature of 58 K.

Solution: $F_{dB} = 10 \log_{10}(1.2) = 0.79$ dB.

Exercise 7.3
Find noise factors and noise figures corresponding to excess temperatures of 100 K, 150 K, 290 K, and 1000 K.

Answers: 10 $\log_{10}(1.344) = 1.29$ dB, 10 $\log_{10}(1.517) = 1.81$ dB, 10 $\log_{10}(2) = 3.01$ dB, and 10 $\log_{10}(4.45) = 6.46$ dB.

7.3 NOISE TEMPERATURE AND NOISE FACTOR OF CASCADED AMPLIFIERS

In Figure 7.2 we show an incomplete schematic of a cascade of stages of amplification with differing amounts of gain and the different amounts of excess noise they add. It must be noted that excess temperature is related *only to the input of each amplifier stage*. Therefore, the noise after two stages of amplification is

$$N_{out_2} = G_1 G_2 \, kB\left(T_s + T_{e_1}\right) + G_2 \, kBT_{e_2},$$

and after the third stage,

$$N_{out_3} = G_3 \left[G_1 \, G_2 \, kB\left(T_s + T_{e_1}\right) + G_2 \, kBT_{e_2} \right] + G_3 \, kBT_{e_3}.$$

Normally, the requirements in communications systems would be in terms of output signal-to-noise ratio. In terms of the relations we have obtained,

$$\left(\frac{S}{N}\right)_{out} = \left(\frac{G_1 G_2 G_3 S_i}{G_3\left[G_1 G_2 \, kB\left(T_s + T_{e_1}\right) + G_2 kBT_{e_2} \right] + G_3 kBT_{e_3}} \right)$$

$$= \left(\frac{S_i}{kT_s B}\right) \frac{1}{1 + T_{e_1}/T_s + T_{e_2}/G_1 T_s + T_{e_3}/G_1 G_2 T_s}.$$

The first term in the last relation is the input signal-to-noise ratio. Being added in the denominator, we can see how the excess temperatures of the various stages work to reduce the output signal-to-noise ratio. Of special interest is the way that gains in the first stages tend to reduce the effects of excess temperature in the later stages. This is not the form of equation usually seen for cascades, but it is *always correct, regardless of source temperature T_s.* If, and only if, it should happen that the source is at standard temperature, we can write the noise factor for the cascade shown as

$$F = F_1 + \frac{F_2 - 1}{G_1} + \frac{F - 1}{G_1 G_2}.$$

If more stages need to be accounted for, one can continue the pattern with each successive term being the next noise factor minus 1 and divided by one more stage gain.

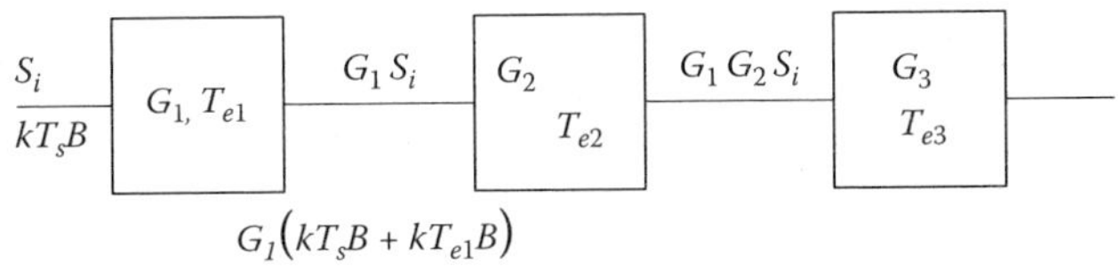

$$G_1\left(kT_s B + kT_{e1} B\right)$$

FIGURE 7.2 Signal and apparent noise present between stages of amplifier cascade.

In the expression in terms of temperatures, each successive excess temperature gets divided by one more gain.

Example 7.6

Suppose the first stage has a power gain of 50 and excess temperature of 87 K, the second stage has a gain of 100 and temperature of 290 K, and third has excess temperature of 580 K. How much will the input signal-to-noise ratio be reduced? Assume the source is at standard temperature $T_0 = 290$ K.

Solution: Taking just the second fraction in the excess temperature equation, we have

$$\frac{1}{1+\dfrac{87}{290}+\dfrac{290}{50\times290}+\dfrac{580}{50\times100\times290}} = \frac{1}{1+.30+.02+.0004} = 0.757,$$

which is the amount by which the input signal-to-noise ratio is reduced.

Exercise 7.4

This exercise will demonstrate how very important it is to have good gain in the early stages of an amplifier for good noise performance. Suppose that in the amplifier above, the only change is that the first stage power gain is only 5. Now find the reduction in the output signal-to-noise ratio.

Answer: 0.649.

7.3.1 NOISE FACTOR OF ATTENUATING CABLE

As was demonstrated in Chapter 1, transmission lines such as coaxial cables all have a characteristic impedance such that if that impedance is used as a load impedance, the input impedance at the other end will be equal to the characteristic impedance, and hence the line is matched at both ends if we let it be driven by a generator also having the characteristic impedance. The result of this is that the load is correct to obtain available noise power whether it is connected as a load or connected directly to the generator; the available noise power is the same at either place. The cable may be considered to be at room temperature, so we can say its noise figure is input S/N divided by output S/N. So we have

$$F = \frac{(S/N)_{in}}{(S/N)_{out}} = \frac{S_{in}}{S_{out}}$$

because the noise powers are the same. However, we have

$$S_{out} = KS_{in},$$

where K is a number less than one, and the noise factor is $1/K$.

This can be a double dose of bad news because this cable can be the first *amplifier stage* that a high-frequency signal passes through, and it not only has a noise figure to worry about, but it also has a gain less than one to divide into other terms. We will see this problem in our examples.

Example 7.7

A high-frequency receiver having a noise figure of 4 dB is driven by a cable having 3.01 dB of attenuation. What is the resulting noise factor and figure of the cascade?

Solution: We must first convert all noise figures into noise *factors*:

$$3.01 = 10 \log_{10} F; F = 10^{3.01/10} = 2.$$

This means that the power gain of the first amplifier in the chain is 1/2:

$$F_2 = 10^{4/10} = 2.51,$$

so, overall,

$$F = 2 + \frac{2.51 - 1}{0.5} = 5.02 \text{ and } F_{dB} = 10 \log_{10} 5.02 = 7.007 \text{ dB}.$$

Exercise 7.5

Rework the example above with a cable that has only 1.0 dB of attenuation.

Answer: 5.0 dB. So we see that small amounts of attenuation have less drastic effects.

Exercise 7.6

Now, consider a good means of mitigating the effects of attenuation. Suppose that we precede the cable by a low-noise preamplifier having a power gain of 10 and a noise factor of 2.0. Find the overall noise factor and figure for 1 dB and 3.01 dB of loss in the cable, which is now of course amplifier number 2.

Answers: $F = 2.216$ or 3.46 dB for the 1-dB cable; $F = 2.40$ or 3.80 dB for the 3.01-dB cable.

Exercise 7.7

In a cascade of amplifying stages, the first has a noise figure of 1 dB and a power gain of 10 dB, the second stage noise figure is 3.01 dB and the power gain is 20 dB, and the third has a noise figure of 6.02 dB and a gain of 23 dB.

    a. What is the noise figure of the cascade?
    b. What is the excess noise temperature of the cascade?
    c. If this amplifier is fed by a source at room temperature, what must be the input signal-to-noise ratio for an output ratio of 10? (Hint: Remember that the definition of noise factor is $(S/N_{in})/(S/N_{out})$ when the source is at room temperature.)

d. The input signal is 10^{-14} watts or 10 femtowatts. The antenna is aimed at a point in the sky where the noise temperature is 30 K. The amplifier bandwidth is 500 kHz. Find the input S/N and the ratio at the output of the amplifier.

Answers: 1.34 dB; 99 K; 13.62; 48.3, 11.2.

Exercise 7.8
The amplifier in Exercise 7.7 is fed through a cable having a loss of 0.5 dB. Now rework all parts of Exercise 7.7 with the added stage up front.

Answers: 4.23 dB; 478.5 K; 26.49; 48.3, 2.85.

Exercise 7.9
Again, we will fight the cable loss problem by preceding *it* with a preamplifier having a noise figure of 1.0 dB and a gain of 10 dB. Find the noise figure of the newest cascade, its noise temperature, and what the input signal must be for an output S/N of 10, if the source temperature is still 30 K.

Answers: 1.178 dB, 90.4 K, 8.3 f.

7.4 TRANSISTOR GAIN AND NOISE CIRCLES

When one gets to actually designing a low-noise amplifier, it is often found that the operating point for maximum gain may be very different from that for a minimum noise figure. However, there exists a technique for finding circles upon which the gain or the noise figure is a constant, which can be plotted on a Smith chart. Then one can choose an operating point that is a good compromise between the gain of the stage and its noise performance. Let us again consider the MRF571 transistor. In Section 4.5, we found the input reflection coefficient for a simultaneous conjugate match as $G_{GM} = 0.890\angle{-178.71°}$. However, the data sheet for the transistor says that at this collector current $I_c = 5.0$ mA, and at a frequency of 1.0 Ghz, the minimum noise figure of 1.5 dB is obtained for a reflection coefficient connected to the input of $G_G = 0.48\angle148°$. The sheet also gives a resistance needed to compute the noise circles, $R_n = 7.5$W. The procedure is that for a succession of noise figures higher than the optimum, one computes an intermediate variable N, where

$$N = \frac{F - F_{\min}}{4R_n/Z_0}\left|1 + \Gamma_{opt}\right|^2 .$$

Then the noise circles are centered at $C_N = G_{opt}/(1 + N)$ and have radii

$$R_N = \frac{\sqrt{N\left(N + 1 - \left|\Gamma_{opt}\right|^2\right)}}{1 + N} .$$

TABLE 7.1
Noise Figures

F_{dB}	F	N	$\left|G_{opt}\right|/(1+N)$	R_N
1.6 dB	1.445	0.0309	0.466	0.154
1.7	1.479	0.0625	0.452	0.215
1.8	1.514	0.0949	0.438	0.262
1.9	1.549	0.128	0.426	0.300
2.0	1.585	0.162	0.413	0.334

Let us tabulate all of this for noise figures increasing from F_{opt} by tenths of a dB (Table 7.1).

As for gain, if one uses G_{GM} at the input, the available gain is 14.0 dB. Let us find circles for increments of –0.5 dB in gain; one calculates a gain factor called g_p, which is an actual gain divided by the square of the magnitude of S_{21}. The centers of the gain circles will be at

$$C_p = \frac{g_p C_1^*}{1 + g_p \left(\left|S_{11}\right|^2 - \left|\Delta\right|^2 \right)}.$$

The radii of the circles will be

$$r_p = \frac{\sqrt{1 - 2K\left|S_{21}S_{12}\right|g_p + \left|S_{21}S_{12}\right|^2 g_p^2}}{1 + g_p \left(\left|S_{11}\right|^2 - \left|\Delta\right|^2 \right)}.$$

Noting that the denominator is the same in both expressions, we tabulate it as part of our systematic progress toward these circles (Table 7.2).

These circles have been graphed onto a Smith chart in Figure 7.3. These circles demonstrate that one will always need to make compromises between the gain and noise figures. If one demands a low noise figure of 1.5 dB, one has to settle for less

TABLE 7.2
Gain Circles

dB Gain	g_p	Denom	Center	Radius
13.5 dB	2.487	1.900	0.834	0.127
13.0	2.217	1.802	0.785	0.190
12.5	1.976	1.715	0.734	0.246
12.0	1.761	1.637	0.686	0.299
11.5	1.569	1.568	0.638	0.350

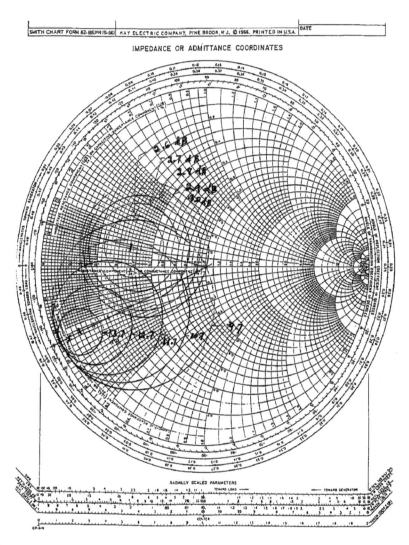

FIGURE 7.3   Gain and noise circles, for example, in Section 7.4.

than the minimum gain graphed, or roughly 11.3 dB. For the lowest noise figure graphed, the best gain available is 13.0 dB. Two circles that just touch each other are for a noise figure of 1.7 dB and a gain of 12.5 dB. How, the reader may ask, does one make a choice among the various possibilities? The answer is that one must look for the best overall performance based upon the noise factor of the total *following* stages. The following example illustrates how this might be done.

Example 7.8

Assume that the overall noise figure of the following stages is 3.0 dB. Let us consider the case where $F_1 = 1.5$ dB and the gain of the first stage is 11.3 dB. What is the resulting overall noise figure?

TABLE 7.3
Answers for Exercise 7.10

G_1	F_1	F for $F_2 \to$ 3.0 dB	5.0 dB	7.0 dB
11.3 dB	1.5 dB $\to$ 1.4125	1.486	1.573	1.710
12.5 dB	1.7 dB	1.535	1.600	1.705
13.0 dB	2.0 dB	1.598	1.656	1.745

Solution:

$F_1 = 10^{0.15} = 1.4125$
$G_1 = 10^{1.13} = 13.489$
$F_2 = 10^{0.300} = 1.995$
$F = 1.4125 + 0.995/13.489 = 1.486$

Exercise 7.10
In order to really see the circumstances when it is important to emphasize first stage gain and when it is preferable to have a minimum first stage noise factor, the reader is asked to compute the overall noise figure for the three operating points mentioned, for several values of following noise factor. For convenience in spelling out the problem, the tabulation in Table 7.3 is given as a start, and the reader is directed to try for the answers in the latter three columns.

Generalizing these results, we see that when the later stages have a low noise factor, the best overall noise factor occurs at the lowest value for the first stage. However, as one gets higher noise factors in the later stages, it becomes preferable to have higher gain in the first stage even though its noise factor is compromised.

Exercise 7.11
Considering the widespread nature of noise problems, it is not surprising that the data for drawing noise circles are scarce. We will borrow some data from one of the earliest application notes published by Hewlett-Packard. This was perhaps the first widespread promotion of the application of scattering parameters to optimize design and assess the results for high-frequency transistors.

The data, at 4 GHz, were

$$S_{11} = 0.552\angle 169°$$

$$S_{12} = 0.049\angle 23°$$

$$S_{21} = 1.681\angle 26°$$

$$S_{22} = 0.839\angle -67°$$

$$F_{min} = 2.5 \text{ dB}$$

$$\Gamma_{opt} = 0.475\angle 166°$$

$$R_o = 3.5 \ \Omega.$$

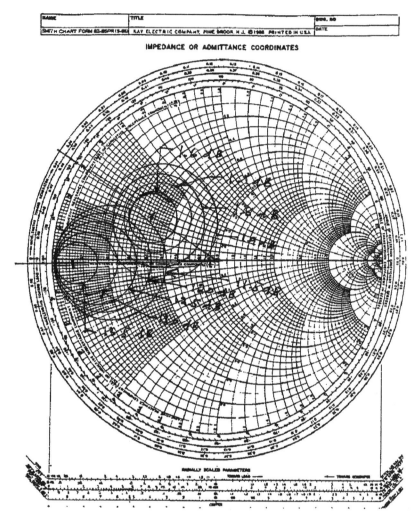

FIGURE 7.4   Gain and noise circles for Exercise 7.11.

Answers: $|\Delta| = 0.419$, $K = 1.012$, so the transistor is unconditionally stable at this frequency. Figure 7.4 is a copy of the author's solutions for the noise and gain circles for the transistor.

7.5   NARROWBAND NOISE

It was previously established for most of the frequency range presently used for communications that the noise one must deal with is "white," which means that the density of the power in terms of bandwidth is *uniform*. Indeed, if we accept that the available noise power is kTB, where T is the system noise temperature and B is now the *effective noise bandwidth*, then we can quote the noise power spectral density as kT. In order to understand the effects of high-frequency noise upon the output noise of a receiver, it may be convenient to consider the noise power in a narrow band as being that of a sinusoid of variable amplitude and phase, having the frequency of the middle of the band. If one doubts the experimental validity of such an assumption, one is invited to feed the voltage from a white noise generator into a low-pass filter and to feed the filter output to a locked-in oscilloscope. The result is apt to contain *very stable zero-crossings with spacings related to the filter bandpass*, albeit a signal that bounces around greatly in amplitude and thus may be interpreted as varying in amplitude and phase. Let us examine the implications of this in an example.

Example 7.9

Suppose we have an unmodulated carrier at 1 Mhz and a passband 10 kHz wide, centered at 1 Mhz. Say we want to split the passband into 10 equal smaller bands. The bands would be 1000 Hz wide. Hence, as shown in Figure 7.5, our approximation would be to show equal noise spectral entries centered at 995.5, 996.5, 997.5, 998.5, 999.5, 1000.5, 1001.5, 1002.5, 1003.5, and 1004.5 kHz. Now the fact that both the amplitude and the phase are varying may be interpreted as saying that we have, for example, varying amplitudes of *both*

$$\cos(2\pi \times 1.0005 \times 10^6 t) \text{ and } \sin(2\pi \times 1.0005 \times 10^6 t).$$

Consider the effects of the noise on the modulated envelope of the carrier. We may expand the cosine term as

$$\cos(2\pi \times 10^6 t)\cos(2\pi \times 500 t) - \sin(2\pi \times 10^6 t)\sin(2\pi \times 500 t).$$

Now, let us add to this a *large* carrier term, for example, $100 \cos(2\pi \times 10^6 t)$. The resulting envelope is the square root of the sum of the squares of the coefficients of the two carrier frequency terms, specifically

$$E(t) = \sqrt{(100 + \cos(1000\pi t))^2 + \sin^2(1000\pi t)}$$

$$= \sqrt{10^4 + 200\cos(1000\pi t) + \cos^2(1000\pi t) + \sin^2(1000\pi t)}.$$

If we add cosine squared and sine squared of anything, we get unity every time. However, in our present situation, that adds only 0.01% to the 10,000 up front, and may be ignored. If we then factor the 10,000 out of the radical, what we have left is

$$E(t) \cong 100\sqrt{1 + 0.02\cos(1000\pi t)}.$$

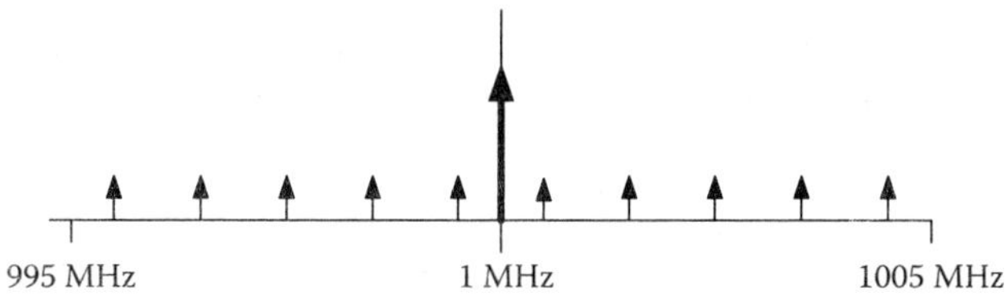

995 MHz   1 MHz   1005 MHz

FIGURE 7.5   Carrier plus phasor representation of narrowband noise.

Applying the binomial theorem, we say the square root of 1 + pretty small is approximately 1 + (pretty small)/2, so the effect on the envelope from that sideband is 100 + cos (1000πt). The noise sideband retained an identity; a sideband 500 Hz from the carrier affects the envelope as noise centered at 500 Hz, and would have that identity if the noisy signal is passed through an envelope detector. The spectrum shown would contribute another 500-Hz term from the 999.5-kHz component and two 1500-, 2500-, 3500-, and 4500-Hz components. Admittedly, our figure gives a very oversimplified view of the noise that is received. Perhaps the reader would feel more satisfied to picture a forest of noise components separated by 1 Hz.

In the example above, there is another oversimplification that must be dealt with. A principle related to the equipartition of energy in statistical mechanics decrees that, in addition to the noise sidebands listed, there will also be *sines* of 1000 ± 0.5, 1.5, and so forth. Expanding the term corresponding to the example above and adding to the large carrier term, one arrives at Example 7.10.

Example 7.10
The sideband may be expanded as

$$\sin(2\pi \times 10^6 t)\cos(1000\pi t) + \cos(2\pi \times 10^6 t)\sin(1000\pi t).$$

The paired sideband *below* the carrier contributes the same thing except that the second term is negative and cancels the second term of the upper sideband. Now, the resulting envelope simply contains

$$E(t) = \sqrt{10^4 + 4(\sin^2(1000\pi t) + \cos^2(1000\pi t))} = \sqrt{10004} \cong 100.02.$$

Stated in words, the sidebands containing the *sine* terms contributes nothing to the envelope. Hence, one-half of the available noise power is ignored by the envelope detector.

There is an unfortunate circumstance with the envelope detector; there *must* be a strong carrier present for envelope detection to work at all. Carrier power does indeed count in determining whether the receiver is above threshold or not. However, *the presence of the strong carrier contributes nothing to the signal output.* This is a reason why it may be necessary to calculate modulation efficiency in AM systems. The output signal-to-noise ratio, which was doubled by detection, will be reduced proportionally to modulation efficiency. Let us illustrate this with an example.

Example 7.11
An AM system with a modulation efficiency of 40% is right at threshold. What will be its signal-to-noise ratio after envelope detection?

Solution: Threshold means an S/N of 10. This is doubled to 20 by the detection process but reduced in proportion to the efficiency. The net result is detected:

$$S/N = 10 \times 2 \times 0.4 = 8.0.$$

There is one more circumstance allowing one to increase S/N, even after detection. Suppose that one's receiver had more bandwidth than was needed to handle the signal. It may not be convenient at all to reduce the receiver bandwidth to the minimum value needed. There is still the possibility of using a low-pass filter on the detected signal. Because the extra noise maintained its identity and spectrum, it can be filtered out by the so-called postdetection filter, a low-pass filter that passes the signal intact but discards the higher frequency noise. This becomes very important in FM; best performance is found for $\beta \gg 1$. Carson then decrees that to handle this signal, one needs much more bandwidth than simply twice the maximum modulation frequency. The Carson bandwidth must be used in the threshold calculation. However, the postdetection filter can deliver quite an acceptable output S/N in FM systems, even if the signal were exactly at threshold at the receiver input.

Example 7.12

Suppose that in the example above we know that our receiver had 50% more bandwidth than was needed to pass the signal. What would be the signal-to-noise ratio if one used a postdetection filter with exactly the bandwidth needed to pass the signal?

Solution: We simply divide out the receiver bandwidth and multiply in the bandwidth of the postdetection low-pass filter. Since noise is in the denominator here, we would have

$$S/N = 12 \times 150\%/100\% = 18.$$

It is important to emphasize that a noise-canceling effect similar to that in envelope detection occurs in synchronous detection. Remember that the synchronous detection process multiplies the high-frequency signal by a carrier, which translates the modulated signal down to baseband and does a similar shift to the noise sidebands; this process again has terms canceling each other in the terms involving the sines. Thus, in synchronous detection, the input S/N is again simply doubled. However, if it happened that one was using more bandwidth than was necessary to pass the signal, a postdetection filter could again strip off the extra noise. Summarizing, then, we may generalize the result after demodulation of amplitude-modulated signals, whether the signal detection is done by synchronous detection or by envelope detection. Both methods successfully ignore 50% of the effective input noise power; therefore, the process of demodulation doubles the input signal-to-noise ratio, no matter which demodulation method is used. We may expect this result whenever the signal is over a threshold defined by the received signal power being at least 10 times noise power. Note that if one is receiving too much noise power to satisfy the threshold requirements, there is no repairing the situation if one is using an envelope detector; when the noise is too strong to satisfy the threshold, it increasingly seizes control of the envelope, so that a large amount of noise is detected. If, however, one has enough signal to satisfy the threshold requirements and is using more bandwidth than the *signal* requires, one can further improve the signal-to-noise ratio by reducing the bandwidth to that required for the signal.

Exercise 7.12

In a synchronously detected DSB-SC system, the received signal-to-noise ratio is 50. However, the receiver has 60% more bandwidth than is needed. Find the output signal-to-noise ratio if an appropriate postdetection filter is used.

Answer: 160.

Exercise 7.13

An AM communication system having a modulation efficiency of 20% is received with a receiver input S/N of 40. However, the receiver has 160% more bandwidth than is needed. Find the output S/N if the appropriate postdetection filter is used.

Answer: 206.

7.6 NOISE AND INTERFERENCE IN FM SYSTEMS

It is interesting that the effects of noise in an FM system can be easily illustrated and understood in terms of an interfering signal that happens to fall in the passband of the FM receiver. Consider Figure 7.6. The specifications we place on what we have illustrated there is that the amplitude of carrier A is strong compared to the amplitude I of the interfering signal. The frequency of the interfering signal is an amount f_I above the carrier. We take steps to combine the signals. We have

$$A\cos(\omega_c t) + I\cos(\omega_c + \omega_I)t.$$

Using the equation for cosine of a sum, we expand the interference term and get

$$A\cos(\omega_c t) + I\cos(\omega_c t)\cos(\omega_I t) - I\sin(\omega_c t)\sin(\omega_I t).$$

If we now combine the $\cos(\omega_c t)$ terms, we get $A + I\cos\omega_I t$. However, our assumption that the carrier term is much larger than the interfering term permits us to neglect the interference in comparison with the carrier term. Now, let us look for a phase-shift term; its tangent will be the negative coefficient of $\sin\omega_c t$ over the coefficient of $\cos(\omega_c t)$, or approximately

$$\theta = \tan^{-1}\left(I\sin(\omega_I t)/A\right) \cong \frac{I}{A}\sin(\omega_I t).$$

A frequency detector will detect instantaneous frequency, which is the time derivative of phase, thus

$$f_i = \frac{1}{2\pi}\frac{d\theta}{dt} = \frac{I}{A}\frac{2\pi f_I}{2\pi}\cos\omega_I t = \frac{If_I}{A}\cos(2\pi f_I t).$$

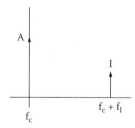

FIGURE 7.6   FM carrier plus interfering signal.

Some interesting characteristics of FM interference are visible here; for one thing, the amplitude is inversely proportional to the carrier amplitude, hence the stronger the carrier, the weaker the interfering signal. This is the source of *quieting*, the phenomenon that reduces the effects of noise as signal strength increases. Another interesting fact is that for an interfering signal separated from the carrier by an amount f_I, the signal out of an FM detector will be *exactly at* f_I. However, perhaps the most difficult circumstance is that the *amplitude* of the interfering signal is also *proportional to* f_I! This can be a great disadvantage. The reader can perhaps remember that when narrowband frequency was studied in connection with AM, it retained its identity at f_I, but its amplitude was *independent of* f_I. Thus, while the noise into the FM receiver is "white," that is, its amplitude was independent of frequency, the noise out of an FM detector has a "parabolic" spectral density, which is to say, noise power density depends upon the *square* of its separation from f_I. If not compensated for, this could make FM a rather noisy medium. The first compensation for FM noise sounds crude indeed; a low-pass filter attenuates signals in inverse proportion to frequency at frequencies far above its corner frequency. Thus, if one places the corner frequency low in the audio range, it will eliminate large amounts of perceived noise. That is the good news; the bad news is that it will also remove so much signal that the claim of high fidelity for FM is a bit ridiculous. However, some unnamed but clever engineer proposed that since one is going to really deemphasize high frequencies after detection, they should be emphasized by an equal amount before they are transmitted. What was settled upon for FM standards was that the low-pass filter should have a time constant of 75 μsec, which corresponds to a corner frequency of 2122 Hz. But before the modulating signal went to the modulator, it was to be fed into a so-called pre-emphasis circuit. This circuit is shown in Figure 7.7(a). The low-pass filter we first spoke of is shown in Figure 7.7(b). The reader should remember that all receivers should have a postdetector filter to get rid of noise corresponding to frequencies higher than one's signal frequencies. Another specification that FM broadcasters have agreed upon is that the highest modulation frequency they will transmit is 15 kHz. Thus, one should follow the FM detector by a low-pass filter with a rather sharp cutoff above 15 kHz. This will allow one to derive very significant benefits from the use of the de-emphasis filter.

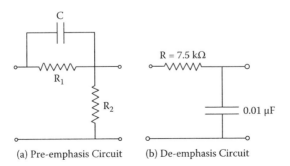

(a) Pre-emphasis Circuit   (b) De-emphasis Circuit

FIGURE 7.7   Pre-emphasis and de-emphasis circuits.

Exercise 7.14

Derive the transfer function for the de-emphasis circuit and specify relations among R_1, R_2, and C, to be an effective pre-emphasizer to work in conjunction with a 75-μsec de-emphasis circuit.

Answers: One needs $R_1 C = 75$ μsec and $R_2 \ll R_1$.

Exercise 7.15

Derive the benefits of standard de-emphasis as follows. Assume that without the de-emphasis filter, one obtains noise power proportional to kf^2 out to 15 kHz, and thus find the noise power one would get without de-emphasis. Then, find the noise power when kf^2 is multiplied by the *power* transfer function

$$H(f) = \frac{1}{1 + (f/f_1)^2} = \frac{f_1^2}{f_1^2 + f^2},$$

where $f_1 = 2122$ Hz. Note that the integral is "improper" in that the power of f in the numerator is no lower than that in the denominator.

Hence, change it to

$$N = k \int_0^B f_1^2 \frac{f^2 + f_1^2 - f_1^2}{f^2 + f_1^2} df = kf_1^2 \int_0^B \left(1 - \frac{f_1^2}{f^2 + f_1^2}\right) df,$$

where $B = 15$ kHz and $f_1 = 2122$ Hz.

Divide the result immediately above into the result without de-emphasis and evaluate the noise improvement for standard FM.

Answer: The ratio of noise without de-emphasis to noise after de-emphasis is

$$\text{Improvement factor} = \frac{(B/f_1)^2}{3\left[1 - (f_1/B)\tan^{-1}(B/f_1)\right]}.$$

For $B = 15$ kHz and $f_1 = 2122$ Hz, the improvement factor is 21.3, corresponding to 13.3 dB.

Exercise 7.16

Suppose the station engineer says, "Our listeners are getting old and losing their high-frequency hearing. Let's cut off our modulation at 10 kHz. What would be the improvement factor if the de-emphasis corner frequency is still 2122 Hz?

Answer: 10.4, or 10.37 dB.

7.7   NOISE IN DIGITAL SYSTEMS

7.7.1   Basic Properties of Gaussian Functions

The effect of noise in digital systems is rather different from analog systems. Instead of a steady irritation overlying the detected analog signal, the noise in a digital system is a problem only to the extent that it causes the detector to make mistakes as to whether a bit received is a one or a zero. Figure 7.8 gives some insight into how this works. Suppose an amplifier had a quiescent point of 4 volts, but there is a rather high noise signal with an AC voltage of 1.0 volt. The Gaussian probability curve shows that the most probable instantaneous voltage will indeed be 4 volts, but there is a small probability of any value between, say, zero and 10. The definitions of probability say that the total area under the curve should be unity, if one goes out to plus and minus infinity. What is perhaps more useful is how much of unity one has failed to include, which is the area under the *Gaussian tail*, as it is sometimes called. The remainder was calculated by the author for a range of special numbers. The quantity is related to the mathematician's *error function* or *complementary error function*, but these definitions contain rather too many 2s for us simple-minded engineers. The common designation of this area under the curve is called the *Q function*. It is tabulated in Table 7.4.

For $v = 3.0$,

$$Q(v) = \frac{1}{\sqrt{2\pi}\,v} e^{-\frac{v^2}{2}}$$

is a good approximation within a few percent.

If one needs to find the inverse, a few sample values are 10^{-5} $Q(4.265)$, $10^{-6} = Q(4.75)$, and $10^{-7} = Q(5.20)$, $10^{-8} = Q(5.61)$.

TABLE 7.4
The Q Function

v	Q(v)	v	Q(v)	v	Q(v)	v	Q(v)
0.1	0.4602	1.1	0.1357	2.1	0.01786	3.1	9.68×10^{-4}
0.2	0.4207	1.2	0.1151	2.2	0.0139	3.2	6.87
0.3	0.3821	1.3	0.0968	2.3	0.01072	3.3	4.83
0.4	0.3446	1.4	0.08076	2.4	0.008197	3.4	3.37
0.5	0.3085	1.5	0.06681	2.5	0.006209	3.5	2.33
0.6	0.2743	1.6	0.0548	2.6	0.004461	3.6	1.59
0.7	0.242	1.7	0.04456	2.7	0.003467	3.7	1.08
0.8	0.2119	1.8	0.03593	2.8	0.002555	3.8	7.23×10^{-5}
0.9	0.1841	1.9	0.02872	2.9	0.001866	3.9	4.81
1.0	0.1587	2.0	00.02275	3.0	0.001350	4.0	3.17

Example 7.13

A noise voltage having Gaussian statistics has zero average voltage and an rms voltage of 1 volt. What is the probability that the instantaneous voltage has an *absolute* value grater than 1 volt?

Solution: The Q function represents the area under *one tail* of the Gaussian curve. We look up $Q(1)$ and read the probability that the instantaneous voltage is greater than +1 volt is 0.1587. Because the Gaussian curve is symmetrical with respect to the mean value of v, there is equal probability that instantaneous voltage is more negative than −1 volt. So the answer to the question above is $2 \times 0.1587 = 0.3174$.

Exercise 7.17

Work the example above for absolute values of 2 volts and 4 volts.

Answers: 0.0450, 6.34×10^{-5}.

Example 7.14

As a further illustration of the Q function, let us suppose that one has a transistor biased at a quiescent collector voltage of 3.0 volts, added to which is an (amplified) noise voltage of 1.0 volts rms. $V_{cc} = 5.0$ volts, and the transistor saturates at 0.1 volt. What is the probability of the noise voltage causing cutoff and saturation of the transistor?

Solution: We have added one more complication of the Q function. If the mean value is not zero, we look up $Q(v - m/\sigma)$, where m is the mean voltage (which here is the quiescent value) and σ is the rms value of the noise voltage. If the number inside the parentheses is negative, we simply look up Q of the positive number, because of the symmetry of the Gaussian curve. To find the probability of the noise cutting off the transistor, we need to evaluate the probability that the instantaneous voltage tries to go greater than 5.0 volts. So, we look up $Q(5 - 3)/1$, getting

$$Q(2.0) = 0.2275; \quad Q(0.1 - 3)/1 = Q(-2.9) = Q(2.9) = 0.001866.$$

Exercise 7.18

Rework the example above, changing only the noise voltage to an rms value of 0.5 volt.

Answers: The probability of cutoff is $Q(4.0) = 3.17 \times 10^{-4}$. To get the probability of saturation, one needs $Q(5.8)$. If one uses the approximate formula, one gets 3.41×10^{-9}. A more accurate table of Q functions gives $Q(5.80) = 3.316 \times 10^{-9}$, so the approximate formula is within 3%.

Exercise 7.19

Now, suppose we have a low-voltage logic system in which the receiver has a system for reading out the ones and zeroes that, at a decision time, if the instantaneous voltage is above 1.1 volts, reads out a logic one. If the instantaneous voltage is below 1.1 volts, a zero is read. (This *decision threshold* would be appropriate if the normal voltage for logic zero were 0 and the normal voltage for logic one were 2.0 volts, and logic zeroes were slightly more probable than ones.) Let the noise voltage be 0.5 volts rms. Find the probability of error for ones and zeroes. (Hint: One uses 0 volts as the mean for logic zeroes, for the probability of errors calculation, and one needs the probability of instantaneous voltage exceeding 1.1 volts. Similarly, one uses 2.0 for the mean for logic ones.)

Answers: P_B (probability of errors) when zeroes are sent is $Q(2.2) = 0.139$; for ones, $P_B = Q(1.8) = 0.033593$.

7.7.2 DETERMINATION OF LOGIC THRESHOLD USING BAYES THEOREM

A mathematician named Bayes derived a theorem to optimize the overall probability of error when ones and zeroes are not equally likely. A little reflection may help the reader realize that one needs to make the more accurate decision on the higher probability digit. Symbolically, the equation for the decision threshold voltage is given by*

$$\frac{z(v_1 - v_0)}{\sigma^2} - \frac{{v_1}^2 - {v_0}^2}{2\sigma^2} = \ln\left(\frac{P(0)}{P(1)}\right).$$

In this expression, the decision threshold is given by z; v_1 and v_2 are the normal voltage levels corresponding, respectively, to ones and zeroes; σ is the rms noise voltage; and $P(0)$ and $P(1)$ are, respectively, the probabilities of zeroes and ones. When ones and zeroes are not equally probable, the overall probability of error is $P_B(\text{overall}) = P_B P(0) + P_B P(1)$, where $P_B(0)$ and $P_B(1)$ are the respective probabilities of error for zeroes and ones.

Example 7.15
Suppose the probability of occurrence of zeroes is 0.7 and that for ones is 0.3. The logic voltage levels are, respectively, 0 and 2 volts. Noise voltage is 0.5 volts rms. Find the optimum decision threshold for the minimum overall error probability and find this overall error probability.

Solution: Substituting in the Bayes relation, we have

$$\frac{z(2-0)}{(0.5)^2} - \frac{2^2 - 0^2}{2(0.5)^2} = \ln\frac{0.7}{0.3} = 0.847 \qquad 8z - 8 = 0.847 \qquad z = \frac{8.847}{8} = 1.106 \text{ volts.}$$

The threshold of 1.100 volt was chosen by guess, but it is seen to be very near optimum. Error probabilities must now be determined by linear interpolation. The author gets $Q(2.212) = 0.01352$ for the error probability for zeroes and $Q(1.788) = 0.0370$. Then, the overall error probability is $0.7 \times 0.01352 + 0.3 \times 0.0370 = 0.0206$.

---

* For derivation, see B. Sklar, *Digital Communications* (Englewood Cliffs, NJ: Prentice Hall, 1988), 740–743.

Had we used the decision threshold of 1.0 volt, which is only optimum when ones and zeroes are equally probable, the error probability would have been $Q(2,0)$ = 0.02275 for ones, zeroes, and overall.

Exercise 7.20
In the example above, suppose that the noise voltage was 0.25 volt rms. For the same probability of ones and zeroes, find the new optimum decision threshold.

Answer: 1.026 volts.

Exercise 7.21
For an rms noise voltage of 0.5 volt, and for logic levels of 0 for zeroes and 2.0 for ones, find the decision threshold for the lowest overall probability of error, if the probability of zeroes is 0.2 and the probability of ones is 0.8.

Answer: The new threshold would be 0.799 volt, and the overall probability of error is 0.0175.

7.7.3 NOISE-CAUSED BIT ERRORS IN COHERENT DIGITAL SYSTEMS

The analysis above would give valid results if the signal were strong enough to provide a detected signal-to-noise ratio that is greater than one for the bandwidth and operating frequency (hence system temperature, including receiver noise and sky temperature) that one is using. The required signal would be much greater than is obtainable in any satellite communication system, and hence a detecting system that effectively increased the signal-to-noise ratio on which the receiver identifies ones and zeroes would be needed. The correlating detector answered the need for what is also called a *matched filter detector*, and the result might be a little startling until one comes to depend upon it. If one is feeding binary PSK (phase shift keying) into a correlation detector, the probability of error for equal ones and zeroes is

$$P_B = Q\left(\sqrt{\frac{2E_b}{N_0}}\right),$$

where E_b is the bit energy and N_0 is the noise spectral density; equal to kT_{sys}, T_{sys} is the system noise temperature, including the effects of receiver noise and the antenna, or "sky" noise, temperature. Bit energy is given by P/R, where P is the power being received and R is the bit rate. There is a slightly disguised factor; since error probability depends inversely on the square root of the bit rate, if one finds an inadequate error rate at a certain bit rate, one only has to demand and design for a lower bit rate. Let us first illustrate this with an example.

Example 7.16
We require an error probability of 10^{-5} or less in a system using binary PSK, at a system temperature of 200 K and a bit rate of 20,000. What power must be received?

Solution: From our Q function data, we observe that $10^{-5} = Q(4.265)$. We massage the error probability for binary PSK a bit, saying

$$(4.265)^2 = 2\frac{E_b}{N_0} = \frac{2P}{RkT_{sys}}.$$

Solving this for power, we find

$$P = (4.265)^2 \times 20,000 \times 1.38 \times 10^{-23} \times 200/2 = 5.02 \times 10^{-16} \text{ watts.}$$

Exercise 7.22
In the example above, suppose one needs an error probability of 10^{-8} or better. What power is required?

Answer: 8.69×10^{-16} watts.

Exercise 7.23
In a binary PSK system having a system temperature of 1500 K and receiving power of 1 femtowatt (10^{-15} watts), what bit rate can be supported for an error probability of 10^{-8}?

Answer: 3870 watts.

Exercise 7.24
Now we can see that with on-off keying being fed into a correlation detector, the fact that a signal is absent for logic zeroes simply takes the "2" out of the bit error expression. For a received power of 10^{-15}, a system temperature of 1500 K, and a bit rate of 5000, what is the probability of error?

Answer: 9.68×10^{-4}.

Exercise 7.25
It turns out that if FSK is detected using a correlation detector, just as with OOK, the probability of error is

$$Q\left(\sqrt{\frac{E_b}{N_0}}\right),$$

assuming that the two frequencies being used are separated by at least the bit rate. Find the bit rate that can be supported for a system temperature of 150 K, a received power of 10 femtowatts, and a required error rate of 10^{-8}.

Answer: 212 kilobits/second.

7.7.4   NOISE IN NONCOHERENTLY DETECTED FSK

The correlation detector is an example of what is called a *coherent detector*. Now, certainly, a mildly adventurous engineer might say, "Let's just move the FSK carrier into the FM band and see what we get as the output of the FM detector." There may also be other methods of noncoherent detection. Let us just bear in mind that by "noncoherent" we do *not* mean irrational or psychotic. Sklar derives the error probability for this case, as

$$P_B = \frac{1}{2} e^{-\frac{E_b}{2N_0}}.$$

Exercise 7.26
Work Exercise 7.25 for a noncoherently detected FSK.

Answer: 184 kbits/sec.

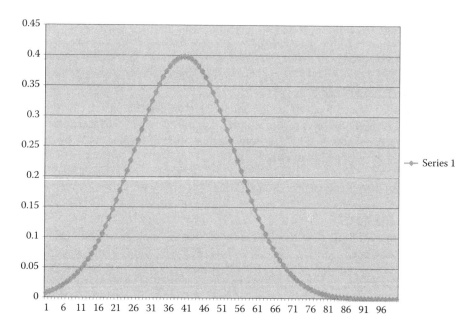

FIGURE 7.8   Gaussian distribution between 0 and 10 v for a mean of 4 v and a standard deviation of 1 v.

8 Antennas and Antenna Systems

It is unfortunate that almost no other subject with which electronics engineers deal seems derived from so much magic and uncertainty. I prefer an approach that seems rational and straightforward to me, putting together building blocks of modest size to arrive at useful results. Under static conditions, the Biot-Savart Law is found to be useful in predicting the magnetic fields that arise because of currents. For high frequencies, our approach will be to use the delayed magnetic vector potential resulting from AC current. From the vector potential, it is a straightforward process to obtain the magnetic field as the curl of the vector potential. Then, electric fields may be obtained from the magnetic fields using the differential form of Ampere's Law. Knowledge of the fields surrounding an antenna may then be used to compute the radiated power.

8.1 FIELDS FROM A CURRENT ELEMENT

Figure 8.1 shows a short wire of length d centered at the origin and pointed along the z-axis, carrying a current for which the phasor representation might be $Ie^{j\omega t}$. Now, the essence of *delayed potential* is that the effects at some point a distance r away from a conductor are simply delayed by the amount of time required for waves traveling at the speed of light to make the one-way trip. Using also some Biot-Savart consequences, the result is that the vector potential has a magnitude and time variation given by

$$\left|\vec{A}\right| = \frac{\mu Id}{4\pi r} e^{j\omega(t-r/c)}.$$

The direction of the vector $\vec{A}$ is parallel to the direction of current flow in the wire. Similar to Biot-Savart, the magnitude is inversely proportional to r, which is the distance from the wire to the point where we want to compute fields, and it is assumed that the medium surrounding the wire is a vacuum, since waves are implied to travel at the velocity of light in a vacuum, which many authors call c. It should be noted that the velocity is only attenuated in the hundredths of one percent if the antenna is in air at standard temperature and pressure. Also, note that no mention is made of the exact value of ϕ at point P; the reason for this is that a symmetry applies here, so the fields do not depend at all upon the specific value of ϕ when the antenna element is right at the origin. However, we must be able to find the fields as a function of the spherical variable θ, so one must dissect the vector component A_z into r and θ components. Use of standard trigonometric relations yields

$$A_r = A_z \cos\theta \text{ and } A_\theta = A_z \sin\theta.$$

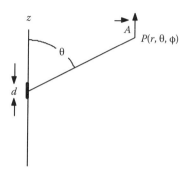

FIGURE 8.1   Vector potential from current element at origin.

Now, of course, the vector $\vec{B} = \nabla \times \vec{A}$.

However, since Ampere's Law, which we use next, contains instead the vector $\vec{H}$, we simply divide the μ out of the vector potential expression, and write

$$\vec{H} = \frac{Id}{4\pi r} \, (\vec{a}_r \cos\,\theta - \vec{a}_\theta \sin\,\theta).$$

The curl expression in spherical coordinates contains, in the r and θ components, only derivatives of the ϕ components, of which we have none, or partial derivatives with respect to ϕ, where there is no ϕ-variation of any components. This only leaves the ϕ-component of curl, given by

$$H_f = \frac{1}{r} \left(\frac{\partial(rH_\theta)}{\partial r} - \frac{\partial H_r}{\partial\theta} \right).$$

In taking the partial derivatives, we must remain aware that there is r-variation in the exponent. Hence, the result, with the $e^{j\omega t}$ time variation understood, is

$$H_f = \frac{Id}{4\pi} e^{-j2pr/\lambda} \sin\,\theta \left(\frac{j2\pi}{\lambda r} + \frac{1}{r^2} \right).$$

If we assume that our medium has zero conductivity, so that there is no conduction component of current and zero free charge, and hence there is also no convection component, Ampere's Law simply relates the curl of $\vec{H}$ to the displacement current density $e \, \partial E/\partial t$. Once one has taken the curl, one integrates with respect to t and obtains the complete expressions for components of $\vec{E}$:

$$E_r = \frac{Id}{4\pi} \sqrt{\frac{\mu}{\varepsilon}} \, e^{-j2pr/\lambda} \cos\theta \left(\frac{1}{r^2} - \frac{j\lambda}{2\pi r^3} \right)$$

$$E_f = \frac{Id}{4\pi} \sqrt{\frac{\mu}{\varepsilon}} \, e^{-j2pr/\lambda} \sin\theta \left(\frac{j2\pi}{\lambda r} + \frac{1}{r^2} + \frac{-j\lambda}{2\pi r^3} \right).$$

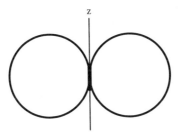

FIGURE 8.2   Vertical pattern of short dipole.

All of these components may be seen to contain several powers of r; however, if one compares the various magnitudes as a function of r, one finds that for $r \gg l$, the dominant component by far is the one that varies as $1/r$. Since these components dominate far from the current element, they are the main determinants of the radiation of energy by the antenna, and are called the *radiation fields*. (The other fields dominate *near* the antenna, hence account for the energy stored near the antenna, and therefore determine the stored energy in the reactance appearing in the equivalent circuit for the antenna. Determining such reactance is a very labor-intensive process, hard to do accurately, and hence will not be done here.) The form of the radiation fields makes it very easy to obtain the total energy radiated. The radiation fields only contain H_ϕ and E_θ; the angular variation of both of these components is $\sin\theta$. Sometimes, one plots the magnitude of a radial vector, which is proportional to $\sin\theta$; this is called the *vertical* radiation pattern* for the antenna and is shown in Figure 8.2. We should also note that the implication of the word *dipole* is that, short though it is, the wire is split in the middle and fed the opposing phases of a balanced (probably 300 ohm) transmission line.

8.2   RADIATED POWER AND RADIATION RESISTANCE

We can obtain the radiated power by integrating the Poynting vector $\vec{E} \times \vec{H}$ over a sphere of radius r. We saw above that the fields that matter in terms of radiated power are E_ϕ and H_ϕ, and since the fields are perpendicular to each other, their cross-product is simply $E_\phi H_\phi$. The vectors are also in phase, so we neglect the phase factors and write

$$E_\theta H_\phi = \sqrt{\frac{\mu}{\varepsilon}} \left(\frac{Id}{2\lambda r} \sin\theta \right)^2 .$$

We must remember and use that in spherical coordinates, the differential surface area is $r^2 \sin\theta \, d\theta \, d\phi$. Then, at any radius r, these r's cancel out those in the fields and our integral for radiated power becomes

$$P = \int_0^{2\pi} \int_0^{\pi} \left(\frac{Id}{2\lambda} \right)^2 \sqrt{\frac{\mu}{\varepsilon}} \sin^3\theta \, d\theta \, d\phi .$$

---

* Because radiation is independent of ϕ, the *horizontal* pattern is a circle with the antenna at the center.

The reasonably resourceful engineer should not be intimidated about integrating $\sin^3 \theta$. One writes it as $\sin \theta \sin^2 \theta = \sin \theta (1 - \cos^2 \theta)$; recognizing that $-\sin \theta \cos^2 \theta \, d\theta$ is of the form $x^2 \, dx$ if $x = \cos \theta$; thus the integral is $x^3/3$, so we can say that

$$\int_0^\pi \sin^3 \theta d\theta = \left[-\cos \theta + \frac{\cos^3}{3} \right]_0^\pi = 1 - \frac{(-1)^3}{3} = \frac{4}{3}.$$

Next, we can say that since there is no ϕ-variation in the integral,

$$\int_0^{2\pi} d\phi = 2\pi,$$

and total radiated power is

$$\left(\frac{Id}{2\lambda} \right)^2 \sqrt{\frac{\mu}{\varepsilon}} \times \frac{4}{3} \times 2\pi.$$

If the dielectric is a vacuum, the so-called intrinsic impedance

$$\sqrt{\frac{\mu}{\varepsilon}} \approx 120\pi.$$

If I is the peak value of a sinusoidal current, the peak power radiated is $80 \, (I\pi d/\lambda)^2$.

It is often useful to speak of the equivalent load resistance of an antenna, to account for the average power radiated. Thus

$$P_{rad} = I^2 R_{rad}/2,$$

where R_{rad} represents *radiation resistance*; for the "short" dipole,

$$R_{rad} \approx 800 \left(\frac{d}{\lambda} \right)^2.$$

We should probably consider a dipole to be short if it is equal to $0.1 \, \lambda$ or shorter. Even at a length of $0.1 \, \lambda$, $R_{rad} = 8$ ohms would not be a very good impedance match for a 300-ohm line. It has been found that such a short antenna would also have a fairly formidable reactance as part of its equivalent circuit. It works out that if one lets a dipole increase in length until it is nearly one half-wavelength in length, the humongous reactance may completely vanish and the radiation resistance approaches 75 ohms. This may have been the motivation for engineers to desire coaxial cables with 75-ohm nominal impedance, but it turns out to be ill advised because the dipole needs to be fed in such a way that the symmetry plane is at ground potential, and the coaxial line is most satisfactory with its outer conductor grounded, necessitating the use of the transformer, sometimes called a *balun* (meaning a device for connecting a balanced load to an unbalanced line. Clever, no?). One can make a structure that parallels the dipole with another wire that has no gap to connect the generator to* (see Figure 8.3). The reader has probably seen and disrespected a folded dipole antenna if he or she has bought a new stereo tuner or receiver, as one was most likely included in the box. The disrespect may

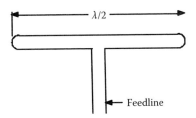

FIGURE 8.3   Folded dipole antenna.

result from the fact that the dipole may be made from the same material as the lead-in wire; to make one at home, one just cuts a length of this wire, called a *twinlead*, with a length about 45% of the wavelength at the center of the FM band, solders the ends together, cuts *one* wire of the folded dipole and connects the lead-in to those new terminals. The author once made measurements on several commercial FM antennas and compared them to a folded dipole mounted on the roof. One sparkly summer night, the dipole was up overnight. It was oriented for the best gain to the southwest, yet was found to provide decent reception from a station over 150 miles to the south-*east*. The folded dipole has 4 times the impedance of the regular dipole, that is, the real part of 300 ohms makes an excellent match to a 300-ohm twinlead. Also, since the balun transformer mentioned above has a 4:1 impedance-transforming, 300-ohm folded dipole fed by a balun is a good match for a 75-ohm coaxial cable. Under some conditions (auto radios and cell phones instantly come to mind) an unbalanced antenna, called a *monopole*, may seem logical. Impedances are thus exactly one-half that for the dipole, so a 1/4 λ monopole will have a radiation resistance around 35 ohms. Another principle we have not mentioned previously is that if the antenna shaft is somewhat "fat" in fractions of a wavelength, say 1% rather than 0.01%, the reactance will not be pronounced and will not have drastic frequency variation.

Exercise 8.1
Find the reflection coefficients with a purely real input impedance of 35 ohms connected to a 50-ohm and to a 75-ohm coaxial line.

Answers: $\Gamma_t = 0.176\angle180°, 0.364\angle180°$.

Exercise 8.2
The last time the author chopped into the lead-in for a car antenna, it looked very much like a low-capacitance cable having 91-ohm characteristic impedance. Suppose 35 ohms are connected to a quarter-wavelength of a 91-ohm cable. What input impedance may be expected at the other end of the cable?

Answer: 237 ohms.

---

* Engineers almost never get criticized for ending a sentence with a preposition. If one is Winston Churchill, it is okay to say, or even roar, "This is a type of errant pedantry up with which I do not intend to put!"

Exercise 8.3

Suppose an auto antenna is telescoped to be 30 inches long. Find the wavelength at 560 kHz (KSFO in San Francisco) and calculate the radiation resistance at this frequency. (Hint: Find R_{rad} for a dipole twice this long and divide by 2.)

Answer: 3.2×10^{-3} ohms. Are you amazed *ever* to pick up an AM station?

8.3   DIRECTIVE GAIN

All practical antennas are directive to some extent. Oftentimes there is expressed a *directive gain*, which may be expressed in several ways. One postulates that it is possible to at least think of an *isotropic radiator*, which is not really possible or even desirable to build, but which would radiate equally in all directions. So in directive gain we say, "Consider an isotropic radiator whose Poynting vector is equal to the maximum Poynting vector of an antenna whose gain is being considered." The gain is then the ratio of the power the isotropic antenna must radiate to have the same maximum Poynting vector divided by the power to the antenna being considered. For simplicity, let us allow the maximum power density (Poynting vector) to be 1.0 watt/meter². Integrating this constant in the θ and φ directions, the isotropic antenna would require 4π watts. Now, since the fields of the short dipole vary as sin θ, the Poynting vector would be $1 \times \sin^2 \theta$. In the previous section, the resulting integral of $\sin^3 \theta$ gave a power requirement of $4\pi/3$ watts. Taking the ratio of 2π watts to $4\pi/3$ watts, we get the directive gain of the short dipole as $3/2 = 1.5$. Sometimes gain is expressed in dB. Since the ratio is one of power,

$$dB = 10 \log_{10} (1.5) = 1.76 \text{ dB}.$$

Allowing the length of a dipole to increase to a half-wavelength, one is rewarded with a pretty modest increase in gain to 1.64, or 2.15 dB. Of course, one does drastically improve impedance-matching possibilities, so that one can send a lot more power out into space.

8.4   ANTENNA ARRAYS FOR INCREASED DIRECTIVE GAIN

A single radiating element will always have a rather limited gain. However, we can consider several antenna elements on a line or in a square in which we can control the amplitude and phase of the driving current. An arrangement in which current amplitude and phase are chosen rather deliberately and stay fixed can have much improved gain. If one has a means of controlling current phases for all elements in real time, the design is called a phased array and may be used to scan the skies, for example, for incoming enemy missiles. Suppose we set out to design an array that will have a fairly high gain in a chosen direction. The general principle is to space individual radiators, which could all be short dipoles, and phase their currents in such a way that the individual elements all reinforce each other in the preferred direction; consider Figure 8.4. We have shown four antenna elements evenly spaced a length *d* along the z-axis. The technical term for this arrangement is a *4-element array*. Suppose we

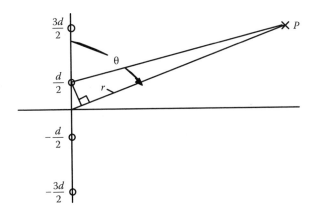

FIGURE 8.4 Antenna array with 4 elements spaced d from nearest others.

wish our beam of radiation to be strongest along the positive z-axis. Let us assume we have chosen the spacing d of the elements to be one-quarter wavelength. Our design scheme then will be to phase the current of each element to be 90° behind the one immediately more in the negative z-direction. The scheme for delayed potential causes the effects of the individual elements to all add in phase as one moves *up* the z-axis. However, as one moves in the negative direction, the element one starts with is already one-quarter period behind the next more negative one, and by the time one moves the distance to the next, the effects are 180° out of phase, so there is cancellation of the effects of pairs of elements in the negative z-direction.

Having designed the array by considering the reinforcement along the z-axis, we must relate to the delayed potential mechanics. Let's say the antenna elements are all short dipoles. For a particular choice of fields, for example, electric fields, we can say that at some distance away, each contribution will have a collection of the same constants, but with individual magnitudes and phases of current and downstairs and r with a subscript representing the distance from that element to the distant point P. If we number the elements from the bottom of the figure, we can write

$$E\text{distant} = \text{constants} \; x$$

$$\left\{ \frac{I_1 \angle \theta_1}{r_1} e^{-j2\pi r_1/\lambda} + \frac{I_2 \angle \theta_2}{r_2} e^{-j2\pi r_2/\lambda} + \frac{I_3 \angle \theta_3}{r_3} e^{-j2\pi r_3/\lambda} + \frac{I_4 \angle \theta_4}{r_4} e^{-j2 \, or_4/\lambda} \right\}.$$

Next, we identify r_1, r_2, and so forth in terms of the coordinates r and θ. We need to keep in mind that Figure 8.4 is a bit inaccurate in implying how close point P is; we showed it on the same page, whereas if we used an accurate scale, it would be many, many pages away. Then, to good accuracy, we can say,

$$r_1 = r + 3d/2 \cos \theta, \; r_2 = r + d/2 \cos \theta, \; r_3 = r - d/2 \cos \theta, \text{ and } r_4 = r + 3d/2 \cos \theta.$$

But the next bit of perspective to keep is that the point P might be 10,000 km away, whereas d might be a meter or less. So the next (very good) approximation is that where r_1, etc., simply appear affecting the magnitude, in the denominator we can simply replace the subscripted r's with the coordinate r. The place where the subscripted values must be carefully used is in the *exponents* representing the phases of the contributions. Let us recall some design choices we made above. Let us make all current amplitudes equal to I. Also, the phase of current 1 will be the reference

$$\theta_1 = 0,\ \theta_2 = -\pi/2,\ \theta_3 = -\pi,\ \text{and}\ \theta_4 = -3\pi/2.$$

We then have four phasors and the phases below:

$$\left\{ 1\angle 0 - \frac{2\pi}{\lambda}\left(r + \frac{3d\cos\theta}{2}\right) + 1\angle -\pi/2 - \frac{2\pi}{\lambda}\left(r + \frac{d\cos\theta}{2}\right) + \right.$$

$$\left. 1\angle -\pi - \frac{2\pi}{\lambda}\left(r - \frac{d\cos\theta}{2}\right) + 1\angle -3\pi/2 - \frac{2\pi}{\lambda}\left(r - \frac{3d\cos\theta}{2}\right) \right\}.$$

If we now factor the cleverly chosen angles out of each term, we can write phases as

$$(-3\pi/4 - 2\pi r/\lambda)$$

$$\left\{ \frac{3\pi}{4} - \frac{3\pi d\cos\theta}{\lambda}, \frac{\pi}{4} - \frac{\pi d\cos\theta}{\lambda}, \frac{-\pi}{4} + \frac{\pi d\cos\theta}{\lambda}, \frac{-3\pi}{4} + \frac{3\pi d\cos\theta}{\lambda} \right\}.$$

Looking at what we have here, we see that the first and last phases are negatives of each other, and likewise the second and third. We recall the identity

$$\cos x = \frac{e^{jx} + e^{-jx}}{2}.$$

We had not previously used our specification that the spacing of the antennas is $d = \lambda/4$. Since we have two pairs of such exponents, we can now write the magnitude of the E-field,

$$|E| = \text{constants} \times I/r \left\{ \cos\left(3\pi/4(1 - \cos\theta)\right) + \cos\left(\pi/4(1 - \cos\theta)\right) \right\}.$$

The quantity inside the {} (brackets) is usually normalized and is then called the *array factor*, which describes how the magnitude varies in comparison to its maximum value. We see that we have attained our design objective, which is to have its maximum at $\theta = 0$. Since $\cos 0 = 1$, both of the cosines are unity at $\theta = 0$, so we can say the array factor is the following:

$$\frac{1}{2}\left(\cos\frac{3\pi}{4}(1 - \cos\theta) + \cos\frac{\pi}{4}(1 - \cos\theta) \right).$$

We have plotted the magnitude of the array factor in Figure 8.5, where the line describes this magnitude, showing simply one display of the E-field; a complete picture would join what is shown with its mirror image. There is quite a broad maximum

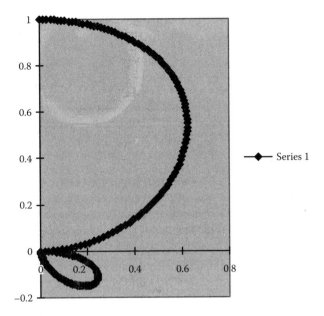

FIGURE 8.5   Radiation pattern of 4-element endfire array.

in the θ = small degrees. Sometimes, we define a *half-power beamwidth*, which here describes the rather blunt cone containing all the values of θ for which the normalized *E* is at least 0.707. By inspection of Figure 8.5, we might estimate that the beam contained all angles out to about 60°. Actually, since we used a spread sheet to compute Figure 8.5, we can read the exact angle as 57°, for a half-power beamwidth of 114°. We also might characterize this array as having uniform field intensity over quite a large portion of space. For a narrower beam, we would need an array with more elements. Suppose we consider adding just one element, keeping the distance between adjacent elements at one-quarter wavelength and the phase difference of the currents adjacent at 90°; when we factor the same exponent out of each term in the array, the angular part factored out would be the exact phase of the centermost element, and there would be no term with which to pair that center term. Hence, the new array factor would have the terms

$$\text{A.F.} = k\left[1 + 2\cos\left((\pi/2)(1 - \cos\theta)\right) + 2\cos\left(\pi(1 - \cos\theta)\right)\right].$$

Again, normalizing so the array factor has a maximum value of unity, it becomes

$$\text{A.F.} = 0.2 + 0.4\left[\cos\left((\pi/2)(1 - \cos\theta)\right) + \cos\left(\pi(1 - \cos\theta)\right)\right].$$

This relation could again be plotted for all values of θ between 0 and 180° to find the new value of the half-power beamwidth; however, engineers who find themselves separated from their computer might get a quick and not too dirty approximation by guessing values of θ and calculating the result. Indeed, as soon as I started this process, I decided to guess the value of 1 − cos θ, starting with 0.3, so that the cosine had to be calculated for 27° and 54°. The complete calculation gives 0.7915 for the result. Thus, we must try two larger angles, one of which is twice the other. Trying 30° and 60° for the angles, we get 0.7464. Trying 32° and 64°, the result is 0.714; 32.5° and 65° gives 0.7064, which is close enough. Now, we say, (working in degrees)

$$32.5 = 90 \times (1\text{-cos } \theta), \theta = 50.3°; \text{ beam width is } 100.6°.$$

Exercise 8.4
How many elements are required for an endfire antenna to have a half-power beamwidth of 75°? Keep spacing at one-quarter wavelength and driving current phases 90° apart. Remember that there is an unpaired element when the number of elements is odd, but they are all paired if the number of elements is even.

Answer: An 8-element array is down only 2.56 dB at 75° beamwidth, a 9-element is down 3.3 dB at 75°.

Exercise 8.5
Check the performance of a 4-element endfire with the elements one-third wavelength apart and phase angles different by 120°.

Answers: Array factor 0.5[cos π(1 − cos θ) + cos(.333 π(1 − cosθ))]; half-power beamwidth is about 98°.

8.4.1 ACTUAL GAIN VIA NUMERICAL INTEGRATION

Just above, we computed beamwidths because it was fairly straightforward once we had determined the array factor. To determine the actual gain, we must integrate the normalized array factor over all possible values of θ to obtain the necessary radiated power. Formally, we will have the integral of a constant in the φ-direction, which gives 2π, but the integral of sin θ times the *square* of the array factor over θ varying from zero to 180°. To do the integral numerically, let's say we will pick increments of θ one degree wide. Hence, we compute the product of the square of the array factor times sin θ at 0.5°, 1.5°, and so forth; multiply by a Δθ of π/180; and add all the contributions. This integral gives 2.0 for the isotropic antenna, so the gain is 2.0 divided by the sum of all contributions. The author did these manipulations for the 4-element array and got a gain of 3.6 or about 5.6 dB.

Exercise 8.6
Do the numerical calculations to obtain gains of the 4-element endfire array with elements one-third wavelength apart and the 8- and 9-element endfires separated one-quarter wavelength.

Answers: Gain of 5.0 or 7.0 dB; gain of 8.0 or 9.0 dB; gain of 9.0 or 9.55 dB.

8.5 BROADSIDE ARRAYS

In the endfire arrays, the design made all the sources interfere constructively, in order to make a maximum of radiation along the line joining the center of all radiating elements. Another conceptually simple idea is to make the sources all combine constructively in a direction *perpendicular* to the line joining the centers of the elements. The term describing this effect is borrowed from naval warfare. If all of a ship's guns are firing perpendicular to the direction the ship is traveling, it is said to be firing a *broadside*. Thus, the antenna array we look at next is called a *broadside array*. Not too surprisingly, it is the result of feeding all the radiating elements *in phase*. Hence, we can say the currents of all elements are at the reference phase, which we can set equal to zero, and there is then no fixed angle to factor out of every term. Suppose we look at an array of eight elements, separated from each other by one-quarter wavelength. The resulting array factor will be

$$\text{A.F.} = 0.25[\cos((\pi\cos\theta)/4) + \cos((3\pi\cos\theta)/4) + \cos((5\pi\cos\theta)/4) + \cos((7\pi\cos\theta)/4)].$$

If we plot the magnitude of this array factor versus the angle θ, we can get Figure 8.6.

There is a caveat or two, as a Latin-speaking politician might say. Perhaps we should say that the pattern represents the antenna's pickup in the direction $\phi = 0$. Except for the confusion factor it might add, perhaps we should superimpose on this picture its mirror image reflected in the vertical axis, because this array still has no way of varying in the ϕ-direction, so the picture should also show the pattern for $\phi = 180°$. Thus, we ought not speak of a *beam width*, because the pattern of a broadside is not really a beam, but more like a Victorian lady's fan that opens up into a complete circle. Still, we can find the gain of this array by again doing a numerical integral. When this is done, we obtain a gain of 4.16, or only about half of what was obtained for the 8-element endfire. Thus, although the range of θ, for which the array factor is 0.7 or greater, is only about 27°, it is uniformly spread over all ϕ, so the gain is much less.

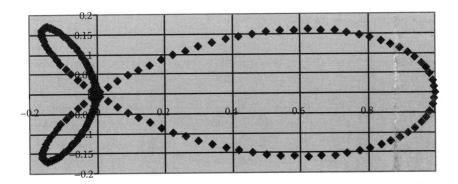

FIGURE 8.6 Pattern of 8-element broadside array.

If the purpose of the antenna array is to produce a beam of radiation in one direction, for example, for a radar system or a deep space communication system, one could consider an array of arrays. Suppose that on the page showing a line of evenly spaced antennas, one would use the same spacing between several parallel lines of antennas. One could continue until there is a square antenna array. If all the currents are in phase, one would have a broadside producing a beam pointing out of the page perpendicularly, with a directive gain depending upon the area of the array. However, if all currents are in phase, this may be a needlessly complicated way of producing a beam. One must be fairly oblivious these days not to have observed the so-called *dish antennas*, which produce a beam by bouncing electromagnetic fields off a reflector with a longitudinal section of a paraboloid. One can characterize the gain by giving the area of the circular opening of the dish. First, however, let us discuss even wire antennas in terms of an "effective area."

8.6 ANTENNA APERTURE AND OTHER APPLICATIONS OF GAIN

We now have the tools for calculating how much power will be picked up by a receiving antenna. We may characterize a receiving antenna by its *aperture*, which is the effective area on the antenna, related to the gain (an antenna's gain is the *same* whether it is used as a transmitting antenna or a receiving one). This effective area is given by

$$A_{eff} = \frac{\lambda^2}{4\pi} \, xgain,$$

where the gain is given as a number (not in decibels).

We assume that each antenna is impedance-matched to its transmission line, and we can write the received power as the product of this area and the Poynting vector at the receiving antenna location. Also, we can obtain the Poynting vector at some distance from the transmitting antenna, by first calculating what it would be at that distance if the transmitted power were spread evenly over a sphere of radius equal to the distance. Then we assume that we have so aligned the antennas so that the transmitting antenna has its maximum gain in the direction of the receiver, and write the Poynting vector as

$$P = \frac{P_t G_t}{4\pi r^2},$$

where P_t is the power being transmitted, G_t is the gain of the transmitting antenna, and r is the distance from the transmitting antenna to the point being considered. Let us look at an example.

Example 8.1

A cellular phone base station has an output power of 100 watts and its antenna array has a gain of 9 dB. Handset antennas may be considered short, with a gain of 1.5.

Assume that the operating frequency is 900 Mhz. Assuming a smooth, flat earth (very unrealistic assumptions, by the way), how far away should the handset be able to receive 10 nanowatts?

Solution: We may write the received power in terms of transmitted power, and both antenna gains and the distance between the antennas, as

$$P_r = \frac{P_t G_t}{4\pi r^2} x A_r = \frac{P_t G_t}{4\pi r^2} x \frac{\lambda^2 G_r}{4\pi} = P_t G_t G_r x \left(\frac{\lambda}{4\pi r}\right)^2.$$

One may use the form of this equation, sometimes called the *range equation*, for which one has data. The reciprocal of the last term, with perhaps the wavelength written in terms of operating frequency and the velocity of light in space, is sometimes called the *free space loss*, and equals

$$\left(\frac{4\pi r f}{c}\right)^2.$$

This author has not come up with a rationale for such terminology, but the reader will not go far in space communications before he/she sees the expression and is now forewarned. Substituting what we have,

$$10^{-8} = 100 \times 10^{9/10} \times 1.5 \times \left(\frac{3 \times 10^8}{4\pi r \times 9 \times 10^8}\right)^2 ;$$

$$r = \frac{1}{12\pi} \times \sqrt{1.2 \times 10^{11}} = \frac{10^5}{\pi\sqrt{12}} = 9189 \text{ m, or about 5.5 miles.}$$

Note that conductivity of the earth, buildings, cars, and other signs of development will most likely greatly reduce this range.

It is not really possible to count the elements on the average TV or other antennas and determine the gain; if one needs the gain, one needs to set up controlled environments and make radiated power measurements. There is just one structure in which fairly accurate estimates can do a good job. Consider the so-called dish antenna, which is basically a paraboloid (the surface generated by rotating a parabola about its axis). One must be sure that the radiating element is located at the focus of the parabola and uniformly "illuminates" the dish. Then, the effective area of the antenna is 0.55 times the area of the circular opening in the dish. If one does the calculation, one sees some truly staggering gains for large satellite dishes.

Example 8.2
Consider a dish having an opening of radius 1.5 m, operated at 6 Ghz. Find the gain of the antenna.

Solution: The effective area is $0.550\pi \times (150 \text{ cm})^2 = 38{,}877 \text{ cm}^2$. The wavelength here is $3 \times 10^{10} \text{ cm/sec}/6 \times 10^9 \text{ Hz} = 5 \text{ cm}$. So we can say $38{,}877 \text{ cm}^2 = (5)^2 \times \text{gain}/4\pi$; gain $= 38{,}877 \times 4\pi/25 = 19{,}542$.

Exercise 8.7
Find the gain of a small dish with a radius of 30 cm, if operated at 15 Ghz.

Answer: 2827.

8.7   THE COMMUNICATIONS LINK BUDGET

Much of the substance of the last few chapters comes together in one calculation having many factors. A transmitter, two antennas, a medium such as a space of incredible extent, and a receiver that adds some noise and decodes the message being sent comprise a communications link. Over the extent of space, digital signaling is the only technique that stands a chance of succeeding. As was demonstrated in the last chapter, the specification that gives the minimum criterion for success is the ratio of bit energy to noise spectral density. One can state the ideal conditions, and then one can, sometimes easily, predict what would go wrong. Early in one of the early moon trips, a video camera was accidentally pointed at the sun, and we earthlings got no live TV thereafter. Both antennas being used are extremely directive, with the result that very small errors in pointing may seriously degrade or completely lose the received signal. Similarly, any polarization preferences of the antennas must be rigorously adhered to, putting a considerable premium especially on the controls of the space vehicle. The power that a spacecraft can transmit depends upon its power supply; if it should happen that the spacecraft is getting its power from the sun, the farther it gets from the sun, the weaker the signal. We can generally expect that space itself is a loss-free medium; however, the distance a signal must travel through Earth's atmosphere can add significant signal attenuation, especially if there is rain or snow in the signal path. Of course, atmospheric attenuation will be worse the farther the signal is forced to travel in the atmosphere. Straight up is of course the best for communication; if the signal path is almost horizontal, one will experience maximum signal attenuation and will also pick up maximum sky noise.

In the previous section we obtained the received signal power in terms of transmitted power, both antenna gains, and free space loss. In Chapter 6, we found we could express the necessary amount of received power in terms of a required value of E_b/N_0, which depended upon the required noise performance. We put all these items plus a margin for error M into the link budget equation, which we write in terms of the transmitter power required:

$$P_t = \frac{(E_b/N_0)_{req} R k T_{sys} L_o L_i}{G_t G_r} M.$$

To save the reader from having to search out the meanings again of all the symbols, one gets a required value for E_b/N_0 from the requirement for bit error performance and the particular from of digital modulation one is using. R is the bit rate in bits per second, k is Boltzmann's constant, T_{sys} is the system temperature (which can depend on many things, such as frequency, how far above the horizon the receiving antenna is pointed, and the receiver noise factor). The factors G_t and G_r are, respectively, the gains of the transmitting and receiving antennas; L_0 is a very large number called *free space loss*; L_i is a factor the author included to account for all forms of "incidental" loss due to incorrect polarization or pointing of antennas; and M is an arbitrary factor the system engineer decided upon to provide a margin for any unexpected, perhaps temporary, sources of signal loss.

Example 8.3
Calculate the transmitter power needed on the moon, about 400,000 km distant, if one requires an E_b/N_0 of 20, a bit rate of 2.0 Mbits/sec, a gain of 4,000 for both transmitting and receiving antennas, a system temperature of 100 K, an incidental loss factor of 0.5, a margin for error of 2.0, and an operating frequency of 4 Ghz.

Solution: For this distance and frequency, the free space loss is

$$L_0 = \left(\frac{4\pi \times 4 \times 10^8 \times 4 \times 10^9}{3 \times 10^8} \right)^2 = 4.49 \times 10^{21}.$$

The transmitted power requirement is

$$P_t = \frac{20 \times 2 \times 10^6 \times 1.38 \times 10^{-23} \times 100 \times 2 \times 4.49 \times 10^{21}}{(4000)^2} = 31 \text{ watts.}$$

Exercise 8.8
NASA's ambitions soared and now they have a little car moving around on the surface of Mars. Its closest distance from Earth is about 6×10^{10} meters. Let us now suppose that the operating frequency is 15 GHz. Payload considerations limit the size of the transmitting antenna to a parabolic dish having an opening of 2 meters in diameter. On Earth, our receiving antenna is a parabolic dish with an opening 20 meters in diameter. We still require $E_b/N_0 = 20$, but we got the system temperature down to 25 K. Transmitted power is 50 watts. Assume incidental losses and margin equal to unity. What bit rate can be supported?

Answers: Free space loss is 1.42×10^{27}, $G_t = 5.43 \times 10^4$, $G_r = 5.43 \times 10^6$, and $R = 1.504 \times 10^5$.

Appendix A

Calculator Alternative to Smith Chart Calculations

There once was a smart-aleck who said, "Blessed are they who go in circles, for they shall be called wheels." The high-frequency engineer might modify that to "for he/she is learning the Smith chart." Before the Smith chart became an inseparable part of the engineer's soul it could be very confusing, and it was very good to have a calculator-based alternative by which to check one's results. Suppose we first look at determining a load impedance from standing wave measurements. While this may seem much more laborious than using one of today's network analyzers, it may be that a new engineer has joined an undercapitalized company that set up standing wave measurement equipment bought at swap meets, for a couple of thousand dollars, rather than a network analyzer costing 20 to 30 times as much. Consider Figure A.1. The slotted line is a jig on which is mounted a section of coaxial cable or waveguide that has in it a longitudinal slot to permit a so-called probe to travel along it. There is a centimeter scale measuring the distance traveled, which is typically numbered from right to left if the equipment was made by the original company of Hewlett and Packard, permitting the numbers to be interpreted as the distance from the load terminals. The probe picks up the total wave voltage on the line, which displays the standing waves; the data one records are the ratio of maximum total voltage on the line to the minimum value (which is called the *standing wave ratio*, VSWR; the word *voltage* was added at the beginning) and the locations of the VSWR minima. A usual first step is to put a high-quality short circuit on the load terminals and determine where the minima are. The short guarantees there is a minimum right at the load terminals, and since standing waves repeat themselves every one-half wavelength, wherever there are minima on the slotted line, we know that such locations are precisely an integer number of half-wavelengths from the load terminals. I like to call these locations values of d_{ref}. Thereafter, when an unknown load is connected, we can say that its impedance is precisely the input impedance presented at the reference values of d. Our calculations must relate the impedance at the locations of the minima when the unknown is connected (I call these locations values of d_{min}) to the value at d_{ref}. Theory tells us that the impedance at a standing wave minimum is purely real and given by the characteristic impedance of the line divided by the standing wave ratio. The equation giving us the numerical calculation of load impedance is

$$Z(d) = Z_0 \frac{1 - |\Gamma| e^{j4\pi(d_{ref} - d_{min})/\lambda}}{1 + |\Gamma| e^{j4\pi(d_{ref} - d_{min})/\lambda}}.$$

Let us do an example illustrating the use of this formula. Suppose, with the short circuit on the load terminals, one found minima (values of d_{ref}) at 10, 12, and 14 cm. Then, with the unknown load connected, one measured a VSWR of 2.5, with minima of 9.6, 11.6, and 13.6. (The envious beginner will note the results of a flawless technique, with well-behaved, precisely spaced data points.) Now, G is the reflection coefficient of the unknown load and is given in terms of the standing wave ratio as

$$|G| = \frac{VSWR - 1}{VSWR + 1};$$

hence, in this example, we get

$$|G| = (2.5 - 1)/(2.5 + 1) = 3/7.$$

The next thing we note is that the distance between adjacent minima on the slotted line is one-half wavelength, or $\lambda/2$. Hence, here $\lambda = 2(12 - 10)$ cm = 4 cm.

Now, we can figure that imaginary exponent of e, as

$$4\pi(10 - 9.6) \text{ cm}/4 \text{ cm} = 0.4\pi \text{ radians.}$$

If one prefers, he or she can think of the exponent as 72°. Hence, one enters one's calculator, which is able to deal glibly with complex numbers, with a complex number having a magnitude of 3/7 = 0.4286 and an angle of 0.4π radians or 72°, depending upon personal preference, subtracts it from $1\angle 0°$, also *adds* it to $1\angle 0°$, divides the former result by the latter, and obtains the normalized load impedance as

$$\frac{Z_L}{Z_0} = 0.7964\angle - 0.7847(\text{radians}) = 0.5635 - j0.5628.$$

Exercise A1.1
When a new unknown impedance is connected to the slotted line above, one observes a VSWR of 3.0, with minima at 10.5, 12.5, and 14.5 cm. What is the unknown normalized load impedance?

Answer: $0.6 + j0.8$.

Exercise A1.2
When yet another unknown load is connected to the slotted line, one measures a VSWR of 2.0, with minima at 9.5, 11.5, and 13.5. Find the load impedance.

Answer: $0.8 - j0.6$.

Index